Breeding Farm Animals

by F.R. Marshall

with an introduction by Jackson Chambers

This work contains material that was originally published in 1911.

This publication is within the Public Domain.

This edition is reprinted for educational purposes
and in accordance with all applicable Federal Laws.

Introduction Copyright 2016 by Jackson Chambers

Self Reliance Books

Get more historic titles on animal and stock breeding, gardening and old fashioned skills by visiting us at:

http://selfreliancebooks.blogspot.com/

Introduction

I am pleased to present yet another title on the Principles of Animal Breeding.

This volume is entitled "Breeding Farm Animals" and was published in 1911.

The work is in the Public Domain and is re-printed here in accordance with Federal Laws.

As with all reprinted books of this age that are intended to perfectly reproduce the original edition, considerable pains and effort had to be undertaken to correct fading and sometimes outright damage to existing proofs of this title. At times, this task is quite monumental, requiring an almost total "rebuilding" of some pages from digital proofs of multiple copies. Despite this, imperfections still sometimes exist in the final proof and may detract from the visual appearance of the text.

I hope you enjoy reading this book as much as I enjoyed making it available to readers again.

Jackson Chambers

This book is dedicated to the memory
of the late Professor John A. Craig
in remembrance of
extraordinary service
to agricultural education
in establishing the teaching
of animal husbandry.

TABLE OF CONTENTS.

INTRODUCTION.

CHAP. I.—EARLIER STOCK BREEDING.

The Arabian Horse—French Horse Breeding—The Thoroughbred—British Stock—Robert Bakewell—Influence of Bakewell's work—Dates of Foundings of the Breeds .... 15—21

CHAP. II.—AMERICAN STOCK BREEDING.

European Stock in America—The American Trotter—Draft Horses from Europe—Coach Horses—Cattle Importations—Advent of Breeds of Sheep—American Breeds of Swine—Hindrances to Stock Raising—Proportion of Registered Stock—Need of Good Stock—State Aid—Federal Aid—Breeders' Opportunities 22—31

CHAP. III.—HEREDITY.

Secrets in Breeding—Indefinite Expressions of Breeders—Results of Experience—Bakewell's Principles—Breeders' Desideratum—Heredity—Heredity has a Physical Basis .... 32—39

CHAP. IV.—FACTS CONCERNING REPRODUCTION.

Female Reproductive Organs—Male Reproductive Organs—Essentials of Conception—Barrenness—Sterility—Number of Services—Size of Litter—Influence of Sire and Dam.. 40—47

CHAP. V.—THE GERM CELLS.

Cell Growth—The Chromatin—The Chromosomes—Preparation of the Germ Cells—Significance of the Chromosomes—Fertilization .... 48—58

CHAP. VI.—THE HEREDITARY MATERIAL.

Scientists and Breeders—Inheritance Through Chromosomes—Possible Chromosome Combinations in Ovum—Possible Chromosome Combinations in Spermatozoa—Equally Probable Results of Fertilization—Why Related Animals Differ—Basis of Controlling Heredity .... 59—64

CHAP. VII.—ORIGIN OF THE HEREDITARY MATERIAL.

The Problem—Necessity and Value of Theory—Darwin's Theory—An Opposite View—Pangenesis—Continuity of Germ Plasm—The Explanations Compared—Relation of Practice to the Theories.................. 65—72

CHAP. VIII.—BREEDING AND SELECTION.

Basis of Control of Heredity—Why Offspring Resembles Parents—Ancestry and Prepotency—How Atavism May Occur .... 73—78

CHAP. IX.—INDIVIDUAL EXCELLENCE IN BREEDING ANIMALS.

Foundation Stock—Types—Value of Type—Need of Full Study—Breeder must be a Judge—Prepotency—Character—Significance of Character—Age and Prepotency—Fancy Points.......................... 79—92

CHAP. X.—PEDIGREES OF BREEDING ANIMALS.

Progeny the Best Test—Form of Pedigree—All Ancestors Must be Studied—Breeding Records of Parents—Similarity of Type in Parents—Value of Show Awards—Fair Estimate of Sire—Advanced Registers—Obscured Merit—Grandparents—Near and Remote Ancestors—Fashion and Family Names—Significance of Breeders' Names—Correctness of Pedigrees...... 93—110

CHAP. XI.—THE OFFSPRING DURING GESTATION.

Relation of Foetus to Dam—Effect Upon Foetus of Maternal Impressions—Need of Care in Management—Nutrition of Offspring—Feeding the Dam—Growth of Bovine Foetus—Influence of a Previous Impregnation. ....111—120

CHAP. XII.—DEVELOPMENT OF YOUNG STOCK.

What Constitutes Environment—Improved Stock for Improved Environment—Feeding Must Support Breeding—Feeding While Young—Good Care Aids Selection—Transmission of Effects of Environment—Biologists on Transmitted Development—Acquired Development in Trotters—Claimed Transmission of Acquired Development—Inheritance Not Related to Sire's Age—How Development Aids Selection—Actual Role of Environment ....121—136

TABLE OF CONTENTS

CHAP. XIII.—DETERMINATION OF SEX.

Influence of Time of Breeding—Influence of Body Conditions—Alternating Ova—Influence of Stronger Parent—Effect of Nutrition—Experimental Evidence—Experiments with Butterflies—The Evidence from Bees—Sex Probably Determined at Conception—The Accessory Chromosome—Significance of the Accessory Chromosomes—Undesirability of Sex Control..137—148

CHAP. XIV.—FOUNDATION AND MANAGEMENT OF A BREEDING BUSINESS.

Breeding an Art—The Breeder's Personal Equipment—Judging Ability—Impartiality Essential—Value of Breed History—Salesmanship—Advertising—Executive Ability—Wealthy Breeders—Location—Home Grown Feeds—Strain More Important Than Breed—Starting from Market Stock—Not How Many, But How Good—Cheap Foundation Stock—Merit in Both Parents Essential—The First Sire—The Second Sire—Phenomenal Sires—Culling the Females—Uniformity in Females—Value of Long Established Herds. ....149—166

CHAP. XV.—INBREEDING AND LINE BREEDING.

Inbreeding Defined—Line Breeding—Opposition to Inbreeding—Thomas Bates and Inbreeding—Barrenness in Early Duchesses—Swine Statistics—Laboratory Experiments—Benefits of Inbreeding—American Hereford Breeding—The Gentry Berkshires—The Principle of Inbreeding—Inbreeding per se—Risk in Out Breeding—When to Inbreed.......................167—187

CHAP. XVI.—MENDEL'S LAW.

Breeding in the Future—Beginnings of New Characters—Mutation—Polydactylous Guinea Pigs—DeVries Experiments—Cause of Mutation—Cross Breeding for New Characters—Mendel—Mendel's Experiments—Mendel's Law—How Mendelian Proportions Occur—Purity of Gametes—Mendelism in Animals—Unit Characters—Application of Mendel's Law—Limitations of Mendelism—Non-Mendelian Characters—Need of Breeder's Records. ....188—208

CHAP. XVII.—BREED RELATIONS.

The Place of Breeds—Evolution of Types—Need for Numerous Breeds—Distribution of Breeds—Community Breeding—Value of Shows—Basis of Awards—Advanced Registration—Registration in the Future—Cross Breeding—Crossing Types—Pure Breeds for Crossing—Limits of Improvement—Effects of Injudicious Breeding—Breeding for Vigor and Prolificacy ....209—221

CHAP. XVIII.—BREEDERS' ASSOCIATIONS.

Origin of Registration—Advantage of Single Registers—Conduct of Herd Books—Relation of Government—Canadian Registration Affairs—Eligibility to Registration—Other Functions of Breed Associations..222—230

CHAP. XIX.—HORSE BREEDING.

Place of the Draft Horse—Cities Set Prices—Earlier Draft Horse Affairs—American Breeding—The Depression—The Revival—Advantages of Foreign Breeders—Stallion Raising—Influence of the Auto-Truck—Breed for top of Market—Draft Types—Market Discrimination—Breeding Carriage Horses—Good Breeding an Essential—Breeding Trotters.........231—247

CHAP. XX.—CATTLE BREEDING.

Influence of the Range—Beef and Valuable Lands—American Progress in Herefords—Evolution of Types—Early Maturity and Size—Advance of Dairying—Advantages of Dairying—Professional and Commercial Breeders—Superiority for Dairy Purposes—Breed Tests—Advances in Dairy Breeding—Form and Function—Extra Influence of Sire—Testing Breeding Cows ....248—265

CHAP. XXI.—SHEEP BREEDING.

Environment for Sheep—Economy of Sheep Raising—English Shepherding—Pedigrees not to be Ignored—Breed Type—Fine Wools in America—Maintaining the Type ....266—277

CHAP. XXII.—SWINE BREEDING.

Need of Improvement—Breed Building—Extremes Needed—Results of Extremes—Show Type—Opening for More Breeds—Mixing Types—Continuity in Farm Breeding—Conservative Breeders—Breeders' Reward ....278—287

PREFACE.

Stock-farming is growing more popular both in theory and in practice. A scientific study of crop production makes clear the necessity of feeding the crops on the farm and the market conditions have afforded encouragement to the breeders and feeders of good stock.

In crop-growing and in stock-feeding, much practical aid has been furnished by the scientists, notably the chemists. The breeders naturally look to the biologists for assistance, but up to the present any directions they may have received have been quite indefinite and not always practical ones.

Notwithstanding the contrary hopes of some earnest and sanguine investigators, it does not seem that the breeding of animals can ever be made an occupation of wholly certain results. Professional breeders of long experience control quite largely the inheritances of their animals. The most that the non-professional or the beginning breeder can hope to accomplish through study is to acquaint himself with the guideposts familiar to those discerning persons who have reached success along the same road.

The main object of this book is to direct attention away from profitless speculations that have necessarily characterized some earlier books, and to stimulate interest in the more tangible, the physical basis of heredity. A scientific study of the physical aspects of heredity leads to conclusions that fully accord with the teachings of the work of our master breeders. It has been the aim to limit discussion to points upon which scientific opinion is quite well agreed, though this has not been altogether possible. Free consultation of the references cited will give a deeper acquaintance with scientific aspects of the question.

I am particularly indebted to Professors F. L. Landacre and H. W. Vaughan for the assistance they have given me. Since this manuscript was prepared "Physiology of Reproduction" has come from the pen of F. H. A. Marshall, D.D. This is a most valuable treatise and contains much to add definiteness to the subjects treated in Chapters IV and V.

F. R. MARSHALL.

Ohio State University.

INTRODUCTION.

It has been said that agriculture is the foundation of all commerce. The records of advances in agriculture parallel quite closely those of national advancement. The ability to adapt the gifts of nature to the needs of man is well exemplified in the changes that have been effected in the domestic animals that play so important a part in the sustenance of the nation.

Domestic animals and their products occupy a very important place in the commerce of the United States. At the close of the first decade of the twentieth century the value of animal products sold from and consumed on farms was equal to one-third of $100 for every memper of the population of the entire country. This has no regard for the large value of work horses used in both city and country. There are numerous reasons for believing that stock raising will hold a much more prominent place in the future than it has in the past in American agriculture.

A contrast of our conditions with those obtaining in older countries shows that a part of our trade is based upon the fact that we still have considerable areas of low-priced lands upon which we rear stock at a cost that permits other nations to buy from us. But with the ultimate occupation of all our lands which are arable, this inequality of conditions affecting production must disappear. It is impossible to fully foresee future conditions, but the opinion seems general and not unreasonable that in considerable part our methods must so change as to result in the production of a large proportion of dairy products. The prosperity of such thickly populated countries as Holland and the Island of Jersey, in which dairy farming is the chief interest, adds much force to this idea, though it must be considered that in those countries a minimum of the labor is performed by hired help. The cornbelt enjoys peculiar natural advantages favoring the growth of its chief crop and it seems likely that for some time to come this section will furnish an important part of the world's supply of meats and of lard.

For every demand for animals for service or as sources of food and clothing there has been produced a special kind of animal. This has also extended to the production of means of recreation and display, as shown in our lighter classes of horses, much of whose service is not immediately connected with trade or ordinary necessities of living. In addition to supplying expressed demands for consumption the work of breeding and adaptation has furnished breeds and families, each characterized by special propensities that peculiarly adapt them to conditions having to do with the economy of production in some particular section or under some particular system.

The group of breeds comprising each of our common classes of stock—horses, cattle, sheep and swine—have sprung from a small number of types of original progenitors rendered somewhat distinct by habits and characteristics which enabled them to survive the rigors and vicissitudes of natural conditions. When viewed as a whole the work of the separation and perfection of so many separate and distinct types from so few natural types appears as a task impossible to human agency even in an indefinite length of time. When we recall that the chief part of this accomplishment covers less than 200 years the marvel grows and an understanding of the principles involved becomes extremely desirable. When we also consider the fact that a large and increasing majority of the agricultural population is chiefly engaged in the production of such animals and that much skill and application are required to maintain the present state of excellence of these numerous breeds, the interest attaching to the principles involved takes on a more practical character. Add to this the further consideration that new breeds are still being added and the older ones are steadily evolving into more specialized and artificial forms, and the estimate of the practical importance of the question is still further heightened.

A SHORT-HORN HERD AT GRASS.

CHAPTER I.

EARLIER STOCK BREEDING.

The first notable achievement in adapting animals to human needs as related to our present-day industry was the development of the Arabian horse. The occupation and manner of living common to the Arabian tribes rendered them dependent upon the fleetness and stamina of their horses. Shorn of all exaggeration and romance which literature has attached to these horses it is undeniable that for their time they were well calculated to be at once the wonder and despair of other peoples. Realizing the great advantage enjoyed in the superiority of their horse, the Arabs very cleverly and wisely surrounded his rearing with an atmosphere of mystery and guarded against his dissemination so as to long retain for themselves the blood which had taken them so long to purify from the coarseness and variableness of its ancestors.

The Arabian Horse.

The peculiar location of France caused its early monarchs to especially interest themselves in the horse stocks of their dominions. Because of the probability of being on unfriendly terms with adjoining countries from which the French soldiery would naturally be horsed it was endeavored to encourage and

French Horse-Breeding.

facilitate the rearing and maintenance of superior horses in order that they might be available for the armies in times of war. Though seriously interrupted at intervals this national assistance to French horse-raising has been continued and was never more efficient nor extended than at the present time. With the exception of some aid to the improvment of fine-wooled sheep, similar aid has not been extended to the other classes of stock. The blood of the Arabian was considerably used at an early date to refine the coarser native stocks, but a principal factor in production of existing types of horses has been the demands for special types of service and the use for breeding of those horses found most suitable to the demands of the prevailing kinds of labor. In later years the Arabian horse has been used but little.

England also drew from the stock of the Arabians in her work of perfecting the light horse for racing purposes, and the work was encouraged by King James I, who imported horses from the Orient in the early part of the seventeenth century. The succeeding reign saw other and more numerous importations of eastern stock, but it was the performance of Eclipse, foaled in 1764 and a great-great-grandson of a horse imported in 1706, which marked the supremacy of the British-bred horses on the race course. The progeny of other horses of the same period showed that, though indebted to the Arabian, the skillful methods of selection practiced by English breeders had produced a much superior animal for their purposes.

The Thoroughbred.

Whether or not early improvers of farm stock profited

by the work of the horse breeders we cannot tell, but it was in the latter part of the eighteenth century that there was inaugurated a most notable improvement of British farm stock. This era of stock improvement resulted in the origination of over a score of separate and distinct useful breeds of horses, cattle, sheep and swine, all well known in America today. American agriculture has drawn mainly on Britain for its live stock. That this is not due to an unreasoned preference for institutions of the mother country is shown by the patronage of continental breeders of Percheron horses, Holstein-Friesian cattle and Rambouillet sheep. Inasmuch as we still import from that small island, the area of which is scarcely equal to that of an average state, considerable numbers of six breeds of cattle, four of horses, nine of sheep and three of swine, the foundation of its animal husbandry is of more than ordinary interest. Although our chief interest is centered in events that transpired subsequent to 1760, it is not necessary to assume that animal husbandry was entirely chaos previous to that time.* British agriculturists of that day appreciated the relation of stock feeding to crop yields. It was recognized that the animals of some counties were quite distinct from those of other counties in their rate and manner of growth and fattening qualities. The necessity of using the best animals as breeders was understood and regarded by some, though it cannot be said that there was anything like a general application of that principle.

British Stock.

---

* For a full review of earliest breeding see Darwin, "Animals and Plants under Domestication," chapter 20.

The development of British breeds of live stock dates from 1760. It was at about this time that Robert Bakewell* assumed the management of the estate on which his father and grandfather had resided at Dishley in Leicestershire.

Robert Bakewell.

Although his most notable success was achieved with the sheep known as Bakewell or Dishley Leicesters, his work in the breeding of Longhorn cattle has been of inestimable value to all branches of the breeding industry. The practices he relied upon in his breeding of Longhorn cattle are still the mainstay of breeders throughout the world.

The accomplishments of Bakewell served his entire country. His surplus stock became distributed through the adjoining counties, but of more importance than this was the force of his example and the spread of information regarding the marked improvement he had effected and his means of attaining his ends. With the need for stock raising becoming more and more apparent and the more discriminating demands of consumers of meats, the example of Bakewell lent an impetus to British stock interests which resulted in that country's reaching the foremost position which she still occupies.

Influence of Bakewell's Work.

The earliest Short-horn breeders, the Colling brothers, were students of Bakewell's and aroused world-wide attention by the prices received at their sale in 1810.

---

*The life and work of Bakewell is well described in an article in the 1894 Report of the "Journal of the Royal Agricultural Society."

On the western side of England the cattle raisers of Herefordshire had produced a class of cattle adapted to their climate and system of raising, but the most effectual improvement was effected by men who were contemporaries of Bakewell or lived after his time.

In 1822 a start was made in recording the pedigrees of Short-horn cattle and a similar work for the Herefords was commenced in 1846. It *Dates of Founding of the Breeds.* was not until 1862 that Scotch breeders of Angus and Galloway cattle provided registration for their cattle though they had attained more than local eminence prior to that date. The Red Polled cattle of Norfolk and Suffolk counties were first recorded in 1874 and the Devons were recorded in 1851.

The improvement of the sheep stock seems to have followed more closely after the work of Bakewell than did that of cattle. It was upon his Leicester sheep that the fame of this great breeder of Dishley chiefly rested and it is not surprising that the shepherds should have been the first to emulate his example of improvement. Although we have a seeming profusion of breeds of British sheep, each was the result of the endeavors of breeders of a particular county to perfect a breed that would be the most economical producer under their conditions of climate and soil and their systems of cropping and feeding. Though there was some use of older breeds in some cases, still the distinguishing characteristics of size and color of face are mainly traceable to similar appearances that happened to be present in the original native stock. Although breeders of Leicesters were working in co-operation prior to the death of Bakewell, the system of registering pedigrees of sheep was not adopted until many years after most of the breeds had been developed and had earned a general popularity in their respective sections.

At an early date Gloucester and parts of adjoining shires has become known for the distinctive characteristics of the sheep native to that section. During the close of the eighteenth and the beginning of the nineteenth century this breed, the Cotswold, received liberal infusion of the blood of the more refined and easy feeding Leicester. Considerable numbers of descendants of Bakewell's flock also found their way into neighboring shires to modify some of the weaker features of the stock that had been developed there.

By the close of the first quarter of the nineteenth century the farmers of the chalky hill lands of the shire of Sussex had, without resort to other blood, brought their sheep to a high order of utility. John Ellman and Jonas Webb did extraordinary service in the perfection of this breed, the Southdown, and though we cannot say what their familiarity with Bakewell's work was, the product of their efforts has been distributed even more widely than the stock reared at Dishley.

In succeeding years the Southdown sheep made a strong impression on those of Hampshire and still later the stock of Hampshire was drawn upon to mate with Cotswolds for the formation of a type with the most useful combination of characteristics for the agriculture of Oxfordshire, and the sheep bearing this latter name were admitted to separate classification at the 1862 show of the Royal Agricultural Society. Nine years previous the same recognition had been accorded the descendants of the native stock of Shropshire and Stafford, though such descendants owed much to the blood of both Leicester and Southdown.

The practical appreciation of the value of carefully bred stock that prompted the formation of so many breeds has never flagged. The limited size of the country and the large population, in spite of importation of food materials, have ensured remunerative prices for animals and their products and the general practice of selling animals rather than crops has sustained the yields from the soil. Practically all animals in every part of the island show a preponderance of ancestry of some of the breeds, and British agriculture is based no less upon the superiority of the farm animals than upon the spirit that would retain or use only the best that could be procured or produced.

CHAPTER II.

AMERICAN STOCK BREEDING.

In America the development of our animal husbandry has afforded a marked contrast to the course of events in other lands. Colonists from various European countries brought with them such stock as was most common to the section from which they emigrated. This gave us horses of Spanish and French blood, cattle and swine of Holland, Germany and contiguous territory. Also there came from Spain the progenitors of much of our stock of fine-wooled sheep. Some of the prominence of British breeds of stock must be attributed to the large number of British colonists, but for the chief part their strength of numbers and popularity has been earned on the basis of utility and adaptability to the requirements of the various agricultural parts of the country.

European Stocks in America.

As early as 1750, Virginia gentlemen brought from England running horses for racing and breeding. There being no place to which they could go for pronounced speed at the trot, the American breeders early began the study and sifting of their horses of mixed blood with a view of perpetuating and intensifying the sources of excellence in trotting speed. Though

The American Trotter.

the Thoroughbred was a prominent factor at the inception of the work and for some time afterward, the accomplishments are entirely accredited to American skill and enterprise.

It was not until the middle of the nineteenth century that the Percheron horse of France made its entrance into America. The history of Shires *Draft Horses* dates from about the same time, *from Europe.* while the Clydesdales invaded the field somewhat later. Our acquaintance with the Belgian is comparatively recent.

In the eighties the present type of the English Hackney obtained a foothold in America and was followed some years later by the French *Coach* and German Coach breeds. Numer- *Horses.* ous horses of carriage type and characteristics have occurred among the trotting stock and the United States Government now maintains a stud for the purpose of so combining the blood of such horses as to perpetuate their carriage qualifications. The Government also supports an attempt to preserve the type of horses descended from the famous Morgan horse of Vermont.

In cattle we had numerous valuable shipments of Short-horns prior to 1820, and influential activities in importing Short-horns into Ohio *Cattle* began in 1833. The seventies saw *Importations.* the attention of British breeders centered upon America and some considerable exportations of Short-horn blood were made to England from America. Importations from abroad, though varying in extent with conditions, have been continuous and are still quite common. From 1875 to 1885 saw the rapid and hard-earned rise and spread of the Herefords and Angus, with the Galloway also making fast friends in sections to which its peculiar virtues commended it. In Herefords we have progressed to the point where we no longer feel the need of recourse to the foreign herds for aid in improvement, though importations of the other breeds mentioned are still common.

Our dairy breeds we have also brought from abroad and it is to the credit of America that her citizens have not been loath to profit by the results of the laborious efforts towards stock improvement in other lands. The breeds imported represent such combinations of characters of adaptability and special usefulness as to allow each part of the country and each class of production to have a breed at least fairly well suited to the peculiarities of the locality or demand.

It cannot be said that the breeds have been distributed, or are even now found, in such surroundings as their founders aimed to serve, but their career is yet so short that natural or reasonable distribution cannot now be expected. Selection and differences in ideals have produced types within breeds with many of the special features common to stock bearing another breed name. We have also established polled varieties of all but one of our imported horned breeds.

Spanish fine-wooled sheep were brought to America before the close of the eighteenth century and from their descendants we have produced an almost confusing number of so-called breeds or strains. These classes of Merinos exhibit as high efficiency in the art of breeding as is evidenced in the productions of any other part of the world. Since 1840 considerable numbers of the Rambouillet sheep of France have been brought in.

Advent of Breeds of Sheep.

The English Leicester was known in America before the Revolutionary War, and several lots of the two other long-wooled breeds arrived subsequent to 1830. By the latter date the Southdown had also earned considerable popularity and the period between 1860 and 1890 saw the establishment and wide distribution of the other down breeds, all of English origin.

Europe furnished America with no breeds of swine bred to the purpose of turning corn into lard, and we have, therefore, three leading American breeds distinguished only by such incidental characters as color and by differences in utility, due to variations in length of standing and the standards of the breeders. These breeds have been descended from imported European stocks which could hardly be said to have been highly improved except in the instance of the Berkshire. In 1835 the Berkshire was established in the section that was to be the birthplace of the Poland-China. This English breed is still imported in small numbers, but the Berkshire of the cornbelt is more useful to that section than is the stocks as bred in England. For what we need of Yorkshires and Tamworths we draw upon England still. It was in 1872 that the National Swine Breeders' Conven-

American Breeds of Swine.

tion adopted the name Poland-China for the hogs that originated in southwestern Ohio and were meeting with much favor. It was several years later that the Chester White was deemed a breed and the Duroc-Jersey received its name from its assembled breeders in 1883. Other breeds of American origin have been produced and retained in restricted areas.

When it is considered that most breeds of live stock have had their residence in America for less than half a century it is no cause for surprise to learn that a large proportion of farm animals bear no evidence of relationship to any breed. Only in the older sections has agriculture taken on anything like a permanent aspect. In such localities depleted soils have emphasized the need of live stock. In a smaller country the dependence of such areas on stock farming would long ago have forced out of existence all but such animals as could show themselves possessed of practical superiorities over all less carefully bred stock. The influx of cheaply raised western stock to supply the population of the East has seriously disturbed the natural progress of agricultural affairs in the more densely peopled states.

Hindrances to Stock Raising.

The western landowners and operators have had an exclusive interest in stock raising and have been in direct touch with the industry, both in the country and at market centers. It is therefore found that in our newer states in which crop-raising has not become general we have a higher average of domestic animals than in places where conditions demanded the best grades of stock but in which that stock could not be produced in competition with the operators of the new, cheap lands of western states.

The advance of each of the breeds toward a higher place in the estimation of the agricultural public has been a steady one but much slower than it should have been. American breeders of note have not been wanting nor has there been an absence of raisers of superior market stock to set an example for their neighbors struggling with inferior work and feeding stock and declining crop yields. The non-agricultural character of a large proportion of the settlers of our lands and their refusal to recognize the need of conserving the fertility of the virgin soils has checked the general adoption of a studied system of stock-farming such as obtains in older, prosperous countries. The existence of other hindering factors such as unsteady values and transportation difficulties must also be recognized.

The native animal exists as a product of proved excellence for withstanding natural conditions. When there is a demand for animals that can utilize and respond to artificial care and feeding and give returns proportionate thereto, the animals produced by artificial selection are appreciated. The improved stock enlarges its domains and adds to the ranks of its devotees not so much through the force of the arguments of its supporters as through the victories won wherever a really good animal is fairly pitted against a native with no inherent possibilities of making response to studied care and feeding.

With the great early appreciation of the imported breeds there was a demand larger than could be fully supplied with such creditable representatives as would be qualified to worst the native. Possession of certificates of registry was too often looked upon as a guarantee of the desired excellence. Many descendants of registered parents had in themselves none of the practical qualifications that the times demanded. The too frequent sales of such stock and the extent to which it was retained for breeding hindered the proper regard of the more meritorious animals and thus, in a measure, those who claimed to be friends of advancement really exerted an influence in the other way. To a considerable extent present-day progress is retarded by the indiscriminate propagation of registered animals, not so much through the injury resulting from their dissemination as by misrepresentation to persons not familiar with the breeds and with what really improved stock actually stands for and can accomplish.

An officer of the Bureau of Animal Industry* estimates that of all horses in the United States 1.02 per cent are registered. For dairy cattle the percentage is given as 1.07, beef cattle 1.05, sheep 0.46 and swine 0.45. It would be of great interest if we could know what proportion of the farm animals of the British Islands are entered in books of record, because there the conquest of the scrub was assured long ago and there is a minimum of animals that are the result of no plan and exhibit no peculiar usefulness.

Proportion of Registered Stock.

Percentages of registration, however, are a crude guide to the status of animal husbandry. Registration figures show the number of animals that have been produced for the express purpose of use as parents of other

---
*Bureau of Animal Industry Report, 1905.

animals which in turn may be designed either for production of still other breeds or for service or slaughter. Many animals of pronounced merit and of carefully selected lineage are never registered and may be superior to some whose lineage is a matter of official record. This applies especially to swine, for although the percentage mentioned is a small one it is well known that but few animals reaching our markets fail to show strong infusions of the blood of the improved types.

Such statistics as are available in a few states show that the majority of the stallions that are siring colts are not of recorded stock. Though this class may include some useful sires it is well known that many, even of those with pedigrees, are not fitted for the service they are allowed to perform.

Actual tests of numerous representative herds of dairy cows in two states show that a large proportion of cows kept are incapable of returning any profit to their owners. One-fourth of the cows kept in one case yield less than one-half the butter fat secured from the better one-fourth of the herd.* It is only necessary to scan the rank and file of offerings of any classes of stock at our market centers to realize that while every section may have some

Need of Good Stock. representatives of breeds resulting from improvement, still much of the stock reared is nearer to the type of the native than to that which the market most highly appreciates. Even though the future should permit the cheap-selling grades to be produced at a profit, it is assured that there will be a more general ap-

---
* Illinois Experiment Station, Circular 106. Indiana Experiment Station, Bulletin 107.

preciation of the higher classes of stock, not only because of their higher market value, but also on the basis of their more ready response to skillful care and feeding.

To encourage horsemen to raise such animals as are most profitable, several states have enacted legislation to prohibit stallions of inferior character standing for public service. It is not the owner of the stallion who is at fault in standing an unsound or low-bred horse so much as it is the fault of the mare owner who elects that the horse he rears shall inherit such inferiority. The stigma placed upon the low-class horse which the state refuses to license is the most effective accomplishment of such laws.

State Aid.

The United States Government allows entrance from other countries, free of duty, of all registered animals intended for breeding purposes. The Government also does special service in some sections to encourage the keeping of better classes of farm animals and, in addition to its endeavors to develop types of horses, it is working with the western sheep interests to produce a type of sheep with the qualifications most needed on the range.

Federal Aid.

It sometimes appears that men have already accomplished most of what can be done in breeding farm animals. Considering the very wide adherence of the majority of stock farmers to unimproved types and the fact that the future will make it imperative that there be reared only such animals as are peculiarly fitted for special purposes, it becomes apparent that the distribution of the products of the breeders' art has only begun. If every unregistered and inferior sire, retaining the grades that are known to be good breeders, could be eliminated from any one of our states, the supply of registered and superior ones would be entirely inadequate to meet the demands.

The foundation of American animal husbandry has been well laid and the work of its perfection is making sure and steady progress, but the extent of past accomplishments is but a fraction of what remains to be done. The average of excellence of stock reared for breeding purposes must be greatly raised. This is to be done largely by the elimination of the undesirable individuals, but the class of showyard merit is certain to be modified to meet the changes needed by the market and by the varying vicissitudes of rearing under different conditions.

Breeders' Opportunities.

As conditions exist today very considerable numbers of American-bred animals are being shipped to other countries and it is impossible to make any reasonable forecast of the extent of such trade in the future. There is need of the services of every one who has the qualifications that enable a man to improve his animals, and every one with the capacity to serve in any branch of the industry is assured of remuneration fully commensurate with what he has to offer.

CHAPTER III.

HEREDITY.

Unusually successful breeders are looked upon by persons not conversant with higher aspects of animal breeding as being possessed of some carefully guarded secrets or rules of mating that give them unusual advantages in their work. Anyone not familiar with the art of breeding cannot appreciate the necessity and efficacy of extended and studious observation combined with careful experience. Obviously there is nothing about an animal's individuality or breeding powers that may not be learned as readily by one person as by another, but the difference lies in the significance of the external indications to variously equipped men, and in their courage and willingness to act upon what they have learned to read in the animals scrutinized. Again it is apparent that by far the chief feature of breeding is in the selection of animals for mating.

Secrets in Breeding.

To formulate any rules or guides from the work and instruction of even the most successful breeders is a very confusing task. Many such men have seriously discussed the teachings of their experience in regard to the relative influence upon the progeny of the male and female parents. In some instances

Indefinite Expressions of Breeders.

we are recommended to select a certain type or class of females to mate with a specified type of male; in others emphasis is laid on ancestors of one sex almost to the exclusion of the other sex. Practically all emphasize the necessity of a good line of ancestry, though just what constitutes such and what weight it should be given in comparison with individual make-up is impossible of clear expression.

A long, careful, practical apprenticeship to the operations of our more capable breeders will bring a person of ordinary natural qualities for the work into an intimacy with the aims and guiding principles of the profession that will remove the seeming vagueness of the precepts of even the best followers of the art. At the same time such a person must still be conscious of a superficiality of his knowledge of the laws and forces with which he deals. It has been said that while the practices of the breeders show much of uniformity in their estimation and application of the principles involved, yet their precepts lack entirely the clearness and similarity of their examples. At first sight this is somewhat discouraging to the student or beginner, and it is the aim of these pages to first look into some facts and conditions that obtain in all breeding and which may constitute a basis for later discussion of principles that govern in all breeding operations.

Both precept and example of all good breeders show a uniformity and fundamental reliance upon the principle commonly expressed in the phrase, "Like begets like." Robert Bakewell could say no more to the eager solicitors of his secret than "Breed the best to the best." A hundred years later the sage of Sittyton with as much frankness and definiteness as was possible to put into words assured those who clamored to know the reason of the unusual fleshing of his Short-horns that "thick-fleshed cattle breed thick-fleshed cattle."

Result of Experience.

The most helpful discussion of Bakewell's ideas that is now available is printed by Youatt* and is herewith quoted:

"Having remarked that domestic animals in general produced others possessing qualities similar to their own, he conceived the idea that he had only to select the most valuable breeds, such as promised to return the greatest emolument to the breeder, and that he should then be able, by careful attention to progressive improvement, to produce a breed whence he could derive a maximum of advantage.

"Under the influence of this excellent notion, he made excursions into different parts of England, in order to inspect the different breeds, and to select those that were best adapted to his purpose, and the most valuable of their kind, and his residence and his early habits disposed him to give the preference to the Long-horn cattle.

"We have no account of the precise principles which guided him, nor of the motives that influenced him in the various selections which he made; but Mr. Marshall, who says that he 'was repeatedly favored with opportunities of making observations on Mr. Bakewell's practice, and with liberal communications from him on all rural subjects,' gives us some clue. He tells us, however, that 'it is not his intention to deal out Mr. Bakewell's private opinions, or even to attempt a recital of his particular

---

*Youatt, "Stock Raisers' Manual," pp. 190-2.

practice.' Mr. Marshall was doubtless influenced by an honorable motive in withholding so much that would have been highly valuable; and we can only regret that he was so situated as to have this motive pressing on his mind.

"He speaks of the general principles of breeding, and when he does this in connection with the name of Bakewell, we shall not be very wrong in concluding that these were the principles by which that great agriculturist was influenced.

" 'The most general principle,' he says (we are referring to his 'Economy of the Midland Counties,' vol. I, p. 297), 'is beauty of form. It is observable, however, that this principle was more closely attended to at the outset of improvement (under an idea in some degree falsely grounded, that the beauty of form and utility are inseparable) than at present, when men who have long been conversant in practice make a distinction between a 'useful sort' and a sort which is merely 'handsome.'

Bakewell's Principles.

"The next principle attended to is a proportion of parts, or what may be called utility of form in distinction from beauty of form; thus the parts which are deemed offal, or which bear an inferior price at market, should be small in proportion to the better parts.

"A third principle of improvement is the texture of the muscular parts, or what is termed flesh, a quality of live stock which, familiar as it may long have been to the butcher and the consumer, had not been sufficiently attended to by breeders, whatever it might have been to graziers. This principle involved the fact that the grain of meat depended wholly on the breed, and not, as has been before considered, on the size of the animal. But the principle which engrossed the greatest share of attention, and which above all others is entitled to the graziers' attention, is fattening quality, or a natural propensity to acquire a state of fatness at an early age, and when in full keep, in a short space of time, a quality which is clearly found to be hereditary.

"Therefore, in Bakewell's opinion, everything depended on breed, and the beauty and utility of the form, the quality of the flesh and the propensity to fatten were, in the offspring, the natural consequence of similar qualities in the parents. His whole attention was centered on these four points; and he never forgot that they were compatible with each other, and might be occasionally found in the same individual.

"Improvement had hitherto been attempted to be produced by selecting females from the native stock of the country, and crossing them with males of an alien breed. Mr. Bakewell's good sense led him to imagine that the object might be better accomplished by uniting the superior branches of the same breed, than by any mixture of foreign ones.

"On this new and judicious principle he started. He purchased two Long-horn heifers from Mr. Webster, and he procured a promising Long-horn bull from Mr. Westmoreland. To these and their progeny he confined himself, coupling them as he thought he could best increase or establish some excellent point, or speedily and effectually remove a faulty one.

"Many years did not pass before his stock was unrivalled for the roundness of its form, the smallness of its bone and its aptitude to acquire external fat; while they were small consumers of food in proportion to their size; but at the same time, their qualities as milkers were very considerably lessened."

Youatt refers to one of Bakewell's bulls to which a few cows were brought at 5 guineas each. He also quotes Marshall regarding Bakewell in the words:

"He likewise gives a curious account of Mr. Bakewell's hall. 'The separate joints and points of each of the more celebrated of his cattle were preserved in pickle, or hung up side by side, showing the thickness of the flesh and external fat on each, and the smallness of the offal. There were also skeletons of the different breeds, that they might be compared with each other, and the comparative difference marked.'"

The following is also taken from Youatt:

"The practice of letting bulls originated in this district, and chiefly with Mr. Bakewell, and was generally adopted. The bulls were sent out in April, or the beginning of May, and were returned in August. The prices varied from 10 to 50 or 60 pounds; but in one case, * * * a bull was let at 80 guineas a season. Further evidence of the estimation in which the Bakewell stock was held is shown in his letting three rams in 1787 for 1,200 guineas."

It is commonly written that Bakewell was very reticent by nature and guarded very closely the "secrets of his operations." It seems more just to consider that to the inquirers of his time the process of selection seemed inadequate, and they found it easier to suppose that there was some carefully guarded factor the possession of which would make them equally successful.

In dealing with the relation of offspring and parents we are in touch with the force commonly spoken of as heredity. Manifestly, the desideratum of the breeding profession is the highest possible measure of control over the force of heredity. The past decade has been marked by unusual progress toward a more comprehensive understanding of the various aspects of heredity. Discoveries have opened new avenues

Breeder's Desideratum.

of investigation; previously puzzling phenomena have been rendered possible of explanation, and there exists a somewhat confident air that in the not remote future the breeding of animals will be placed on a plane of greater definiteness and less uncertainty than it now occupies. To study heredity and its newer and scientific aspects as applied to the former and present conceptions of breeding is the object of the succeeding chapters.

Heredity.

The term heredity is most commonly defined as the tendency of the offspring to resemble the parent form. The same thought has been expressed in "Like begets like." We are reminded on every hand in both plant and animal life that like begets like in a general way so far as species or variety is concerned, and in the main also individuality of offspring is close to that of parent. In an exact sense, however, no animal is a counterpart of either parent. Total merit or separate points as often vary away from as toward what we desire. The making of our breeds has consisted no less in the elimination of the undesirable than in the perpetuation and combination of the better features of such animals as have been regarded as approaching more nearly to the ideal of usefulness and value for the purpose for which bred. Manifestly heredity and its study is as much concerned with a consideration of the minor departures from resemblance of offspring to parental type as with the likenesses.

The idea of heredity finds expression in our common words "heritage" and "inheritance," as implying the transfer of title or possession from one generation to another. The proper study of heredity in animals, however, must not fail to recognize that the young animal's heritage is complete at its birth; no subsequent dependence or connection with either parent for nourishment or protection can be considered as in any sense a hereditary relation; in fact, it will later be shown that hereditary impress was fully conveyed at a much earlier period, namely, at conception.

The confusion resulting from the attempt to apply the truth of "Like begets like" in exact or minute sense arises from the necessity of considering every animal in relation to two parents. The physiology of the reproductive processes having to do with the making of a new animal are well understood and sufficiently easy of explanation to repay careful study by one who would familiarize himself with the fundamentals of the breeder's work. The relation of each of the parents to their progeny and the real ultimate origin is made clear by an understanding of the arrangement and functions of the reproductive organs, more particularly in the female.

Heredity Has a Physical Basis.

CHAPTER IV.

FACTS CONCERNING REPRODUCTION.

The formation of the new animal begins with the union of the material from the male parent with a contribution from the female. Under normal conditions this union takes place within the body of the female shortly after copulation. A general knowledge of the location and construction of the female organs is necessary to a useful understanding of the conditions and processes that have to do with the origin of new individuals. The vulva is the external opening of the female reproductive organs. The vagina is the passage lying immediately inside the

Female Reproductive Organs. vulva and in an ordinary mare is from 8 to 12 inches in length. Three or 4 inches from the exterior opening of the vulva is the opening from the bladder. The os (os uteri) or neck of the womb projects into the forward end of the vagina. Its length is about 2½ to 3 inches and owing to the nature of its walls is ordinarily practically closed except at times of breeding or parturition. The womb or uterus is the part that contains the developing embryo. Its rear end opens into the vagina and its forward part is below and to the rear of the kidneys. In unbred mares the main body of the uterus has a length of from 5 to 8 inches.

The ovaries produce the eggs or the female repro-

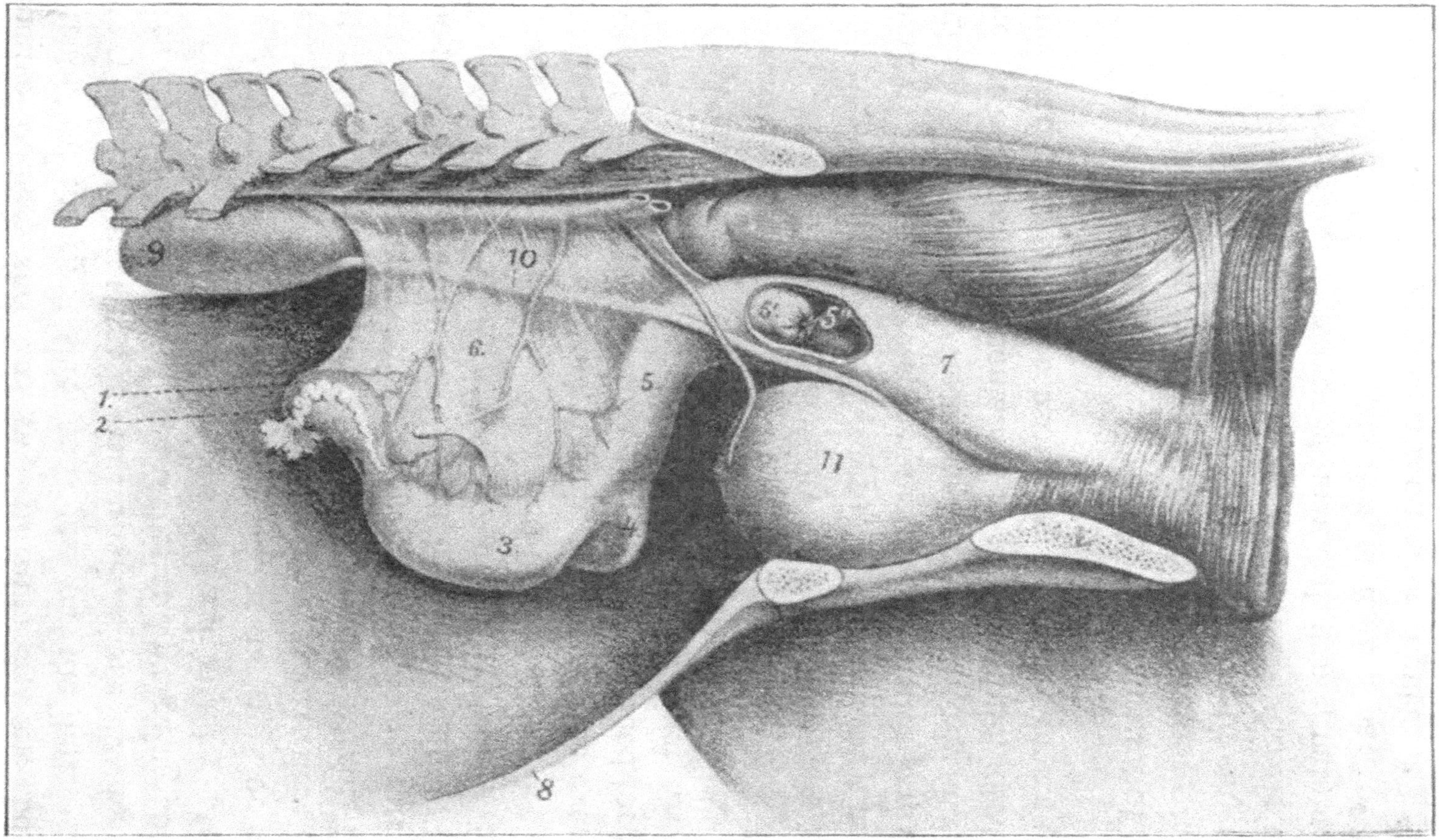

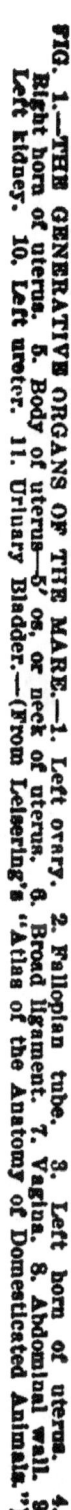

FIG. 1.—THE GENERATIVE ORGANS OF THE MARE.—1. Left ovary. 2. Fallopian tube. 3. Left horn of uterus. 4. Right horn of uterus. 5. Body of uterus—5'. os, or neck of uterus. 6. Broad ligament. 7. Vagina. 8. Abdominal wall. 9. Left kidney. 10. Left ureter. 11. Urinary Bladder.—(From Leisering's "Atlas of the Anatomy of Domesticated Animals.")

ductive bodies, and are two in number, one being situated on the right side and one on the left. The ovary of a young mare is reniform in shape, having its greatest dimension of 3½ to 4 inches and weighing about 4 ounces. The ovary of the cow is much smaller. It is the function of the ovaries to produce the eggs or ova (singular, ovum) from which the new animal develops after a union with another cell from the male parent. The time at which the ovum is ready to meet this body from the male is marked by evidence of being "in heat." The Fallopian tubes connect the ovaries with the womb and through them the ova are conveyed to the latter.

In the male the testicles are analagous to the ovaries of the female. The structure of the parts provided for the introduction of the product of the testicles into the passage of the female is of slight importance in studying heredity and these organs seldom require attention. The testicles produce very large numbers of bodies, the spermatozoa (spermatozoon, singular). These when discharged from the body during the act of service are contained in a white fluid, alkaline in character, the whole constituting the semen or seminal fluid.

Male Reproductive Organs.

After the act of service a considerable amount of the seminal fluid can usually be found upon the floor of the vagina, though the testimony of some of those who have successfully practiced artificial impregnation of mares is that at least a part of the fluid is present in the uterus soon after copulation. The

Essentials of Conception.

spermatozoa, which are capable of some motion, work forward through the uterus into the Fallopian tubes. Here they surround the ovum to the interior of which a single spermatozoon penetrates. This union of the male and female reproductive bodies constitutes fertilization. The united ovum and spermatozoon gravitate to the womb, where, if conditions are favorable, growth and development ensue and conception has been accomplished. It is positively known that a new animal is the result of the union of one ovum and one spermatozoon. The reason for the preparation of such a great number of the male bodies, of which only one is essential to reproduction, lies partly in the likelihood of a large proportion of them failing to reach the ovum, or though arrived there having lost their vitality. Any condition that prevents the union of a healthy ovum with a healthy spermatozoon under normal conditions, renders impossible the production of a new animal. Any such obstruction present in the female is known as barrenness. Inability on the part of the male to supply healthy spermatozoa is spoken of as sterility.

Barrenness. Barrenness may result from a diseased condition of the ovaries. Mares and cows that are continuously in heat and fail to conceive are commonly so affected. In such cases no normal ova are produced and treatment is usually unsatisfactory. Excessive fattening during the growing period may derange the ovaries, especially if the elements that support growth are scantily furnished or if exercise and outdoor life are restricted. The os may be so tightly closed as to prevent the entrance of the spermatozoa. This is common in mares that are quite old when first bred and in heifers kept in very high condition. In artificial impregnation some of the seminal fluid is taken from the floor of the vagina and placed within the womb of the same or another female, thus overcoming any trouble arising from the condition of the os. In difficult parturitions the os is sometimes lacerated and heals with an enlargement that closes the passage. Acidity of the secretions of the womb also causes barrenness. Reproductive cells, both male and female, require an alkaline medium. If through any diseased condition the fluids of the womb become acid, the spermatozoa perish before reaching the ovum or else the fertilized ovum is destroyed. It is for the remedy of such conditions that the yeast treatment is so commonly recommended for uncertain breeders, but it is not uniformly successful.

Sterility.

Absence of procreative power in the male must be due either to failure to produce normal spermatozoa or failure to convey them to the organs of the female. The first named is the most common cause of sterility. The preparation of reproductive bodies is a deep-seated process and draws heavily on the vitality of the animal. It is essential that a breeding animal be maintained in the best of physical condition by judicious and liberal feeding, reasonable exercise and intelligent management. A few causes underlie nearly all manifestations of sterility. In stallions and in aged bulls temporary sterility sometimes follows a slight and often a radical change of location. It is more often caused by an excessive proportion of feeds of a fattening character and by a minimum of work or exercise.

Number of Services. Excessive service may so decrease the number or vitality of the spermatozoa as to produce sterility. In no event can a second service overcome obstacles to conception in either parent. One satisfactory service furnishes a superabundance of spermatozoa; other services can only exhaust the vitality of the male. If wrong conditions are indicated or suspected a remedy should be used before mating is allowed.

It is thought that the breeding of sows late in the period of heat renders more certain the presence of active spermatozoa at the time the last ova leave the ovaries, thus ensuring fertilization of all ova produced. Inasmuch as the number of spermatozoa is so great it is evident that the number of young to a litter must be controlled by the female unless the male be so seriously overtaxed as to lower the number or vitality of spermatozoa, or unless mating occurs at a time too far removed from the time of production of the ova,

Size of Litter.

so that either or both perish before fertilization is accomplished. The number of rudimentary eggs or ova present in the ovary is much greater than the number that can possibly be discharged in a lifetime. Any condition which would augment the production of ova in sheep or swine would of course add to the number of young produced at a birth. A normal, well nourished dam might be ex-expected to mature more ova at one time, but no direct influence can be brought to bear upon this function. The egg cell or ovum is expelled from the follicle of the ovary in which it was prepared at the time of the evidence of "heat" and ordinarily before copulation occurs. It is not known just how long an ovum retains its life after being discharged, but it is probably a considerable time. With the female organs in an entirely normal state it is believed that the spermatozoa may remain active for several days after their introduction. In one case, with a rabbit, spermatozoa are known to have functioned ten days after copulation.*

Influence of Sire and Dam.

The statement was made in the preceding chapter that the new animal develops from two single germ cells, an egg or ovum from the dam and a sperm cell or spermatozoon from the sire. It is natural to entertain the thought that the dam, during the months of gestation in which she carries and nourishes the developing offspring, has opportunity to give to it a stronger impress than was received from the sire. Undoubtedly much depends upon the nourishment afforded the foetus by the dam and this important feature is reserved for later discussion. The actual connection between the membranes enveloping the foetus and the inner surface of the womb allow absorption from the maternal into the foetal circulation of material for the support of growth, but there can be no blood current from one to another. Furthermore, we know that while blood carries building material to the various parts of the body the ability to shape the material rests, not in the blood, but in the contents of the cells that make up the part.

---
*"Transactions Royal Society," Series B, No. 196.

Observation clearly corroborates the idea that the dam has no opportunity to dominate the make-up of the offspring more than is enjoyed by the sire; indeed the claim seems well founded at times that the sire's influence exceeds that of the dam. It is clear that aside from feeding the embryo animal the entire determination of what that embryo is to become resides within the two cells with the union of which the new life was inaugurated, and the one from the sire contains active material equal in amount and determining power to that in the dam's contribution to the embryo offspring. Any particular points or conformation or fitness for special agricultural requirements that are to characterize any descendant of a long line of carefully selected ancestry must have its representation in one of these two sexual cells. Clearly then, it is of first importance that we fully understand the nature and behavior of these germ cells and their relation to the parent body.

CHAPTER V.

THE GERM CELLS.

The spermatozoa were first observed in the product of the male organs in 1677, though their function was not then known. In 1827 the ovum was found and understood to be the seat of new life, though it was not until 1843 that the necessity of the union of ovum and spermatozoa was made clear and not until thirty years later that the significance of such union was realized. Though of unusual shape and make up, each of these reproductive bodies consists of but a single cell. A cell is a unit of structure in all plant or animal tissue as is a brick the unit in a wall. Growth consists of an increase in the number of cells, made possible by the material carried to the growing part by the blood. New cells produced by growth always resemble those existing in the part because they derive their principal and controlling part from the older ones. This controlling part or seat of the greatest activity is the nucleus which is shown at A in the ovum in Fig. 2b. Here the nucleus is small in proportion to the whole cell because of the extraordinary amount of outside material in the egg cell. The contents of the nucleus in a germ cell are believed to be the chief if not the sole vehicle of heredity between the offspring and the parent

Cell Growth.

body within which the germ cell is produced. That the contents of these nucleii of all cells have some unusual qualities is evident from their behavior. It is the practice of the biologist to add clearness to the distinction of parts of material under examination by staining that material with chemical preparations. It invariably happens that when living tissue is so stained the contents of the nucleus take on a deeper and more striking color than

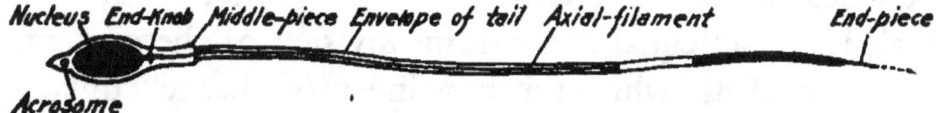

FIG 2A.—DIAGRAM OF FLAGELLATE SPERMATOZOON.

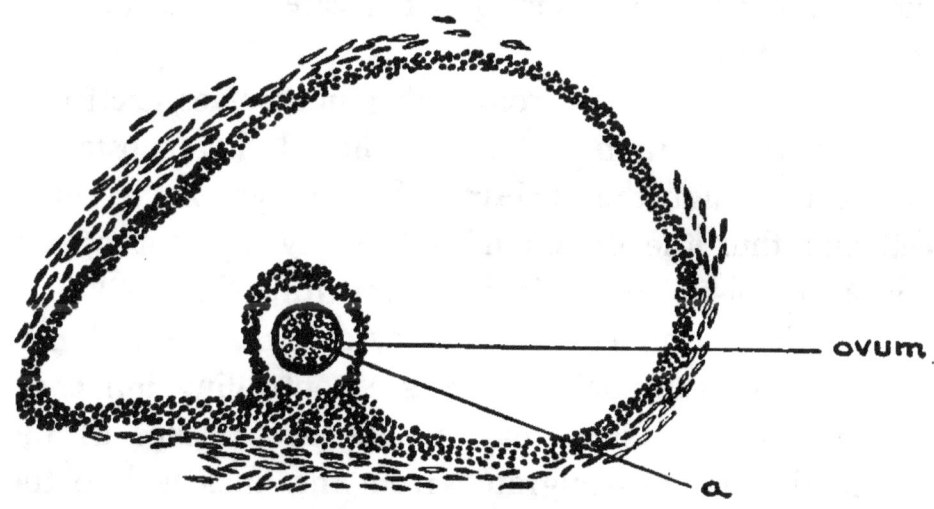

FIG. 2B.—AN OVUM CONTAINED IN THE GRAAFIAN FOLLICLE OF THE OVARY BEFORE BEING DISCHARGED. THE OVUM HAS A DIAMETER OF 1-127TH OF AN INCH. A, NUCLEUS OF OVUM.

do other parts of the cell, evidencing a peculiarity of composition. For this reason the substance within the nucleus is called chromatin. Other grounds for attaching unusual significance to the chromatin are found in the intricate processes provided for in its division every time one cell becomes two cells. While the chromatin is dividing with striking exactness the outer part of a parent cell gives a half of itself to each new nucleus, but this halving evidences little or none of the design and exactness observed in the chromatin division. These facts apply with equal force to all body cells and to germ cells in their preparatory stages. The detail of the processes by which one cell becomes two cells is shown in Fig. 3. The unusual provisions for an equal and careful division of the contents of the nucleus, while the remainder of the cell divides with so little apparent system, lends color to the idea that the nuclear substance is of greatest importance to the resulting cells.

The Chromatin.

After division the chromatin again resolves itself into a granular condition and it is believed that substances pass out through the nuclear wall and control the entire cell and thus the direction of the development of the entire organism resides in the chromatin of its cells. It is sometimes claimed that the cytoplasm, the cell material outside the nucleus, exerts a controlling influence, but evidences of such may be due to presence of the chromatin that has migrated from the nucleus into the cytoplasm. In any body we may trace this chromatin material through successive divisions back to the original ovum and spermatozoon that originated the new being. This chromatin or hereditary material is present in all growing cells in the form of elongated and crudely cylindrical bodies spoken of as *chromosomes*. The number of chromosomes in the nucleii of the cells is the same throughout the body and never varies in the same

THE GERM CELLS

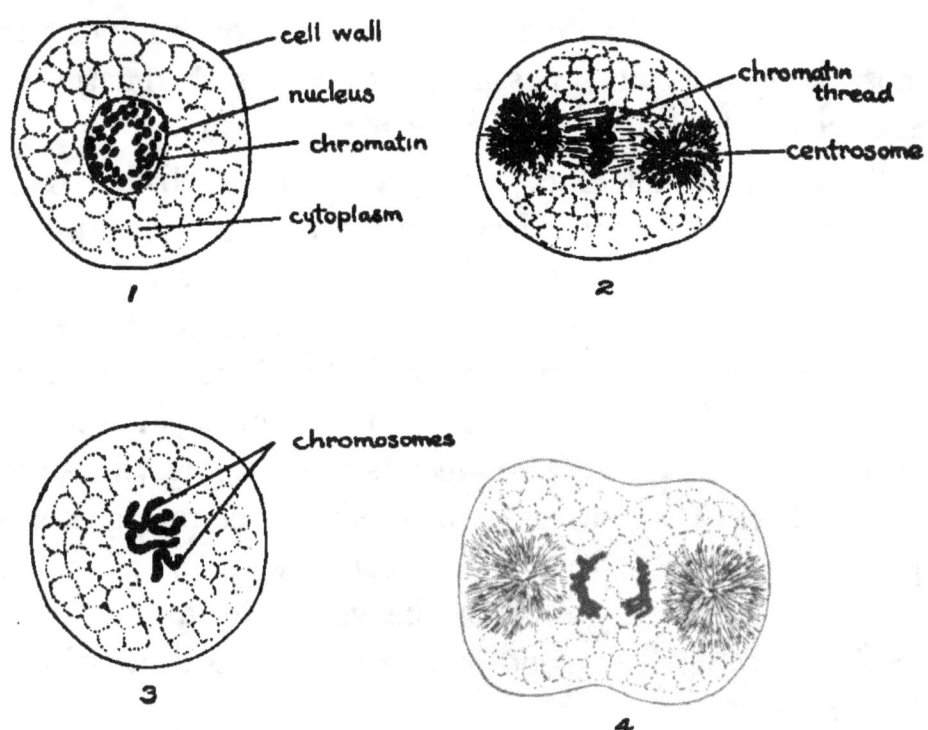

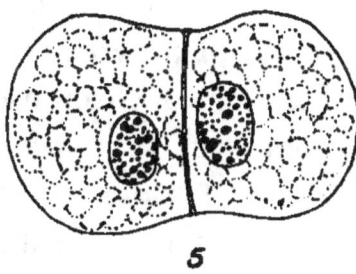

FIG. 3.—PROCESSES OF CELL DIVISION.

1. Cell in resting stage.
2. The chromatin is formed into a skein and held upon numerous strands between two centrosomes.
3. The chromatin has broken up into four chromosomes. This drawing is from an organism that normally has but four chromosomes.
4. Each chromosome has split into two, and the parts are going toward the centrosomes. Note the indentation of the cell wall.
5. Two new cells, each with a nucleus in which the chromatin is in the granular resting stage. (Drawn by H. W. Vaughan.)

species or class of animal. In our common animals the most accurate count possible shows the chromatin in each cell to be made up of sixteen chromosomes.

"The remarkable fact has now been established with high probability that every species of plant or animal has a fixed and characteristic num-
The Chromosomes. ber of chromosomes, which regularly recurs in the division of all its cells, and in all forms arising by sexual reproduction the number is even. Thus, in some of the sharks the number is thirty-six; in certain gasteropods it is thirty-two; in the mouse, the salamander, the trout, the lily, twenty-four; in the worm sagitta, eighteen; in the ox, guinea pig and in man the number is said to be sixteen."*

It was said that the primitive or rudimentary germ cells multiply just as do cells in other parts, namely by each chromosome being split and donating half of its substance to the nucleus of the new cell. If the new animal produced by the union of a germ cell from either parent is to possess the number of chromosomes normal to a cell characteristic of its class some reduction of the number sixteen in the primitive bodies must be effected, otherwise there would be a doubling up and hopeless confusion. The German zoologist, Weismann, in 1887, several years prior to the actual discovery, predicted that it would be found that the first form of the germ cells experienced some such reduction in the number of their chromosomes before reaching their mature form.

This process (maturation) of reducing the number

---
*Wilson, "The Cell," p. 67.

of chromosomes in preparing a mature germ cell was first observed and understood about *Preparation of the* the year 1889. Since that time it *Germ Cells.* has been seen to occur in sections from the ovaries and testicles of most of the larger animals and the process is a common subject of study in zoological laboratories. The special reducing process known in the female as oögenesis, and as spermatogenesis in the male, is apparently solely for taking from each germ cell one-half its chromosomes. No such thing occurs except with cells that are to be used in reproduction. In the male this process is continuous, and perfected spermatozoa are stored in considerable numbers. Maturation or reduction of the chromosomes of the female egg takes place quite rapidly and just prior to union with a spermatozoon. In some instances it is known to have occurred after the spermatozoon has passed through the wall of the ovum. Fig. 4 furnishes a parallel diagram illustrating the formation of spermatozoa and ova. This process of reduction is a basis for explaining many perplexing occurrences in breeding and is worthy of careful examination.

Fig. 5 shows the stages in the reduction of the chromosomes of an egg cell; only six of the chromosomes are shown. In farm animals each germ cell so reproduced would have eight chromosomes. The larger one with which the mass of food is retained is the mature egg or ovum; the other three perish. The processes of reduction of the number of chromosomes follow each other without intervals and are to all appearances solely designed to prevent the doubling of the

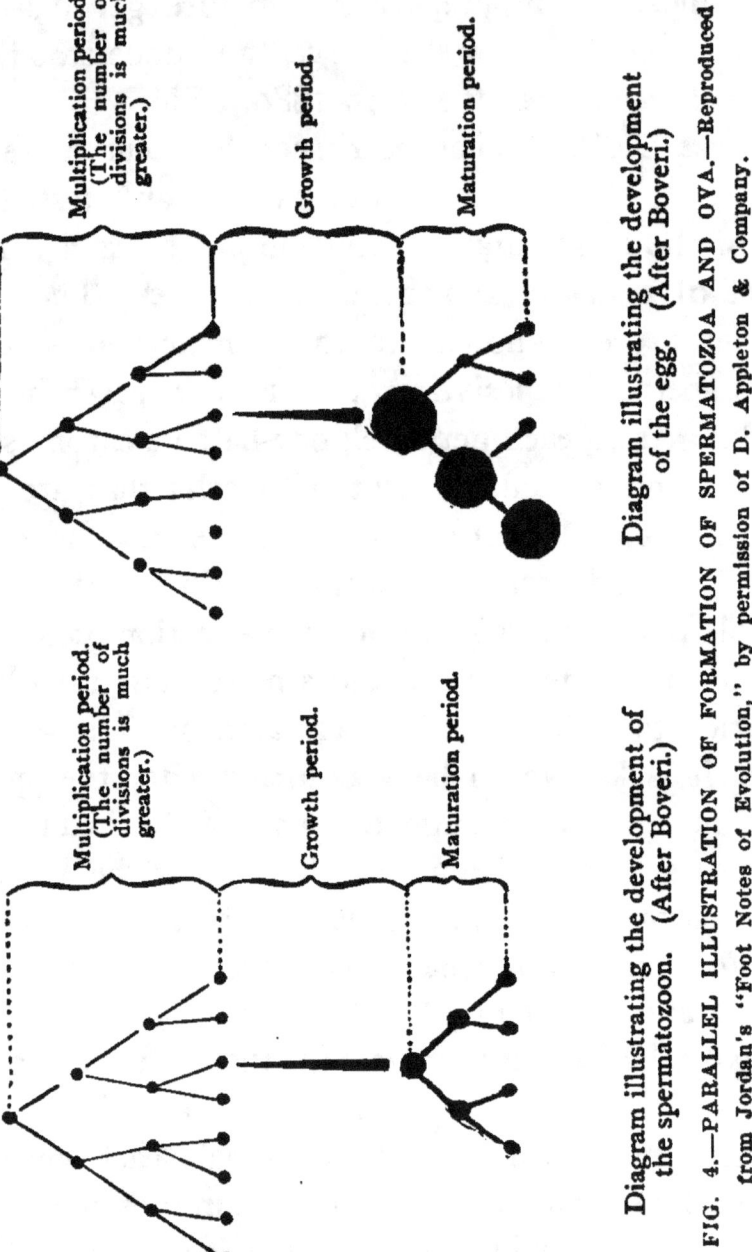

FIG. 4.—PARALLEL ILLUSTRATION OF FORMATION OF SPERMATOZOA AND OVA.—Reproduced from Jordan's "Foot Notes of Evolution," by permission of D. Appleton & Company.

number of chromosomes in the embryo which would follow if reduction did not take place. The later union of spermatozoon and ovum, each with one-half as many chromosomes as are normal to the species, restores the

THE GERM CELLS 55

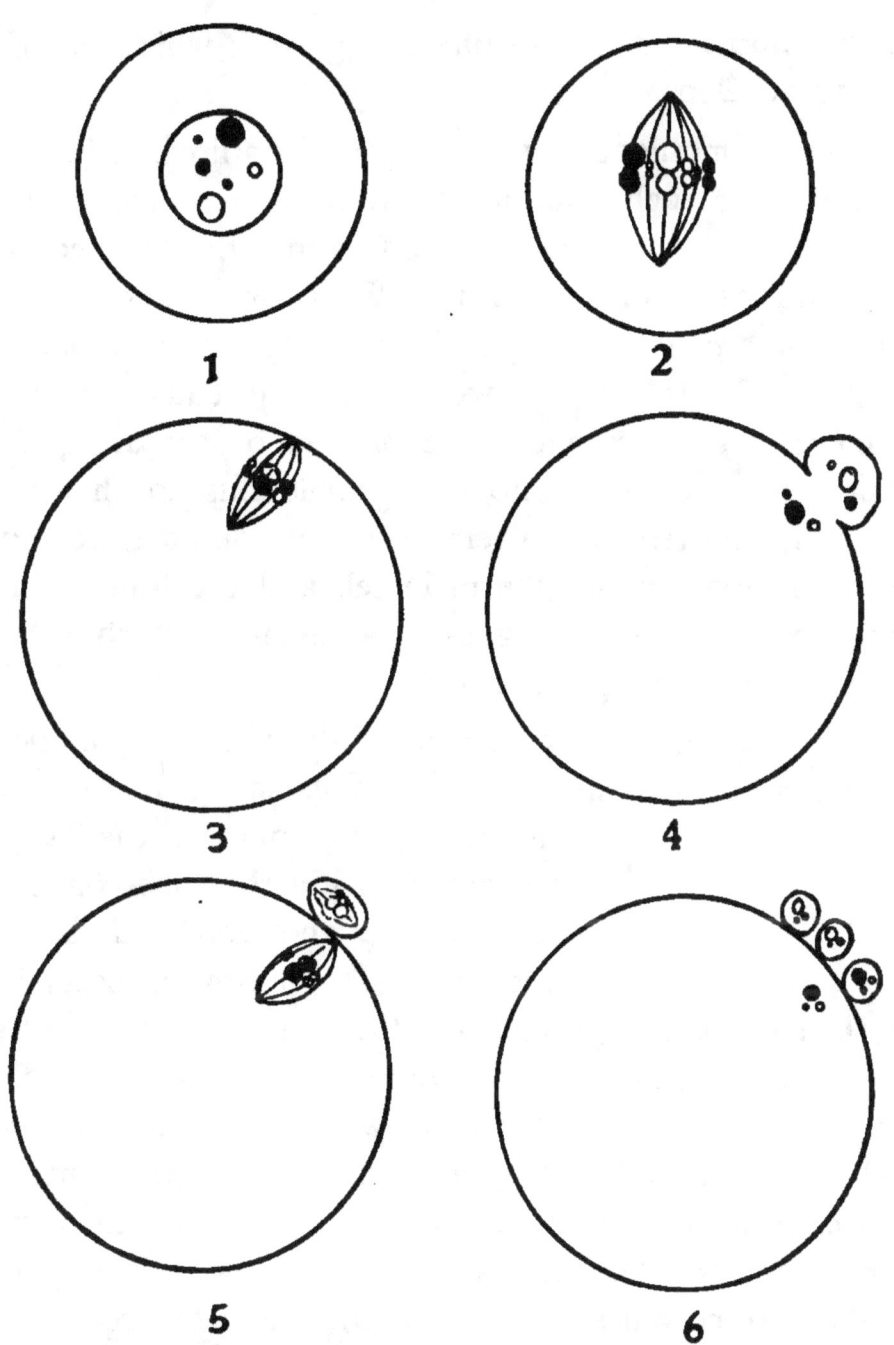

FIG. 5.—REDUCTION OF THE CHROMOSOMES OF AN EGG CELL.—1, early germ cell, oogonium, with whole number of chromosomes—paternal, black dots; maternal, clear rings; 2, division of oogonial cell; 3, first polar spindle; 4, first polar body; 5, second polar spindle and division of second polar body; 6, egg after extension of polar bodies.—Reproduced from "Morgan's Experimental Zoology," by permission of The MacMillan Co.

correct number in the fertilized egg from which the offspring develops.

On the mother's side, then, the new animal is limited to receiving such qualities as were represented in the eight chromosomes that chanced to remain in the ovum. The sixteen originally present in the immature egg were derived in equal numbers from each grandparent. The process of preparing the male germ cells is altogether analogous to that observed in the female. There is no considerable accumulation of food within the male cell and the four bodies produced in the male organs are similar to each other in appearance and possibilities.

Significance of the Chromosome.

The spermatozoa, by virtue of the wriggling motion produced by their tail-like appendages shown in Fig. 2a. find their way to meet the ovum ordinarily within the tube connecting the ovary and womb. It is not known how long a time is occupied by the spermatozoa in reaching the ovum. The difficulty in procuring actual data is obvious. In one case in a rabbit, the ovum and spermatozoa were found united two and three-quarter hours after copulation. Though many spermatozoa attach themselves to the exterior of the ovum but one enters. Thus it is a matter of chance which eight of the chromosomes of the sire will meet the contribution of the dam. Considering the existence of the hereditary material in sixteen unit bodies and allowing for the

*Fertilization.**

---

*The production and union of the germ cells is admirably treated in "Physiology of Reproduction" by F. H. A. Marshall.

THE GERM CELLS 57

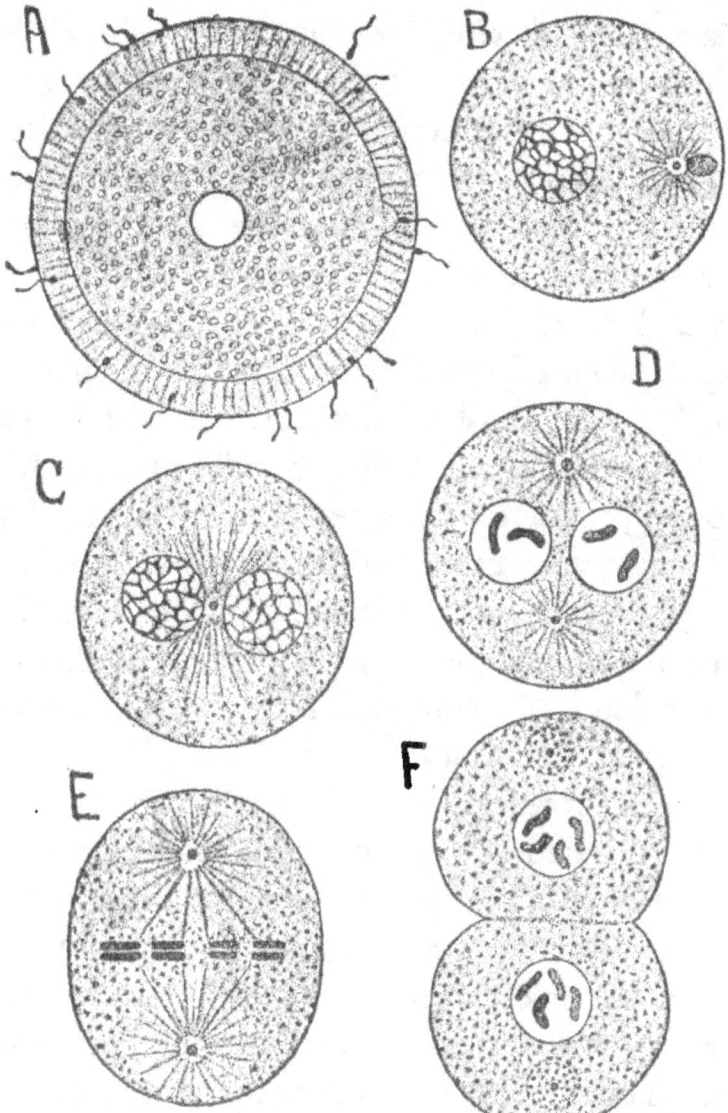

FIG. 6.—THE FERTILIZATION OF THE EGG.—A, egg surrounded by spermatozoa; on the right, one has just penetrated the egg membranes and is entering the egg cytoplasm; egg nucleus in the center. B, egg nucleus with chromatin reticulum on left; on right, the sperm nucleus preceded by its centrosome and attraction sphere. C, egg nucleus on the left, sperm nucleus on the right of the center of the egg; stage immediately preceding the division of the centrosome. D, the centrosome has divided, the two attraction spheres separate to form the first cleavage spindle; the chromosomes of the egg and sperm nuclei clearly visible and indistinguishable (in the figure the egg chromosomes are black, the sperm chromosomes shaded). E, the first cleavage spindle, with splitting of chromosomes. F, completion of first cleavage; two-celled stage; each nucleus contains four chromosomes—two from the egg and two from the sperm. (After Boveri.) Reproduced from Jordan's "Foot-Notes to Evolution," by permission of D. Appleton & Company.

variable tendencies contained in the individual chromosomes there need be no occasion for surprise when successive matings of the same parents fail to produce identical progeny. The spermatozoon having entered the ovum, their chromosomes coalesce and the seed or embryo of the new animal is complete. With favorable conditions growth and development ensue. The repeated divisions carry the leaven of the chromosomes to all parts of the body to form the most minute portions of the new individual. The development of the embryo is quite analogous to the production of a new plant from food secured from the natural sources and built up in the seed.

All of the non-nuclear material present in the fertilized egg is brought there by the ovum, but this material is not considered to convey influences of importance in the subsequent development.

As stock breeders, and therefore interested in the problem of heredity, we are not primarily concerned with the embryonic stages succeeding the union of the ovum and spermatozoon. Viewed in any way the production of a perfect foetus from the enlargements and divisions of two single special cells is a most marvelous process. Though marvelous it is no less comprehensible than is the development of a mature fruit-bearing plant from a single seed. The chromatin or the virtual seed material sends off its various component parts, and representation in the successive stages of change and the chromatin in any cell of the completed form is traceable directly back to the reproductive cells. The further study of this tangible vehicle of heredity is therefore of fundamental interest.

CHAPTER VI.

THE HEREDITARY MATERIAL.

Scientists and Breeders.
Although stock breeders have received many interesting suggestions and directions from scientists during the last few years the facts of heredity are far from being an open book even to the scientists. It may be stated here that the most suggestive expressions the scientists have given to the breeders relate to the creation of new forms and the production and fixation in our domestic animals of characters not now common. In these pages the endeavor is chiefly to present such matter as may permit a more complete understanding of the physical basis of breeding, and the object is not so much the discussion of means of adding new types and characters as it is to stimulate a study that shall result in greater uniformity of excellence among the existing stock and a closer resemblance of the majority to the present best. In arriving at the fertilization of the ovum by the spermatozoon, or the planting of the seed of the new animal, we have adhered to incontrovertible facts. Though entirely probable it has not been fully demonstrated that the chromatin is the exclusive seat of heredity. Even if it is not it is the active part of the cells which do carry all of heredity, and its changes are highly significant.

So far as concerns any intimate knowledge of the make-up of the chromosomes or the distribution among them of the control of various portions of the body we are entirely in the dark. We may at best recognize the combined chromosomes as carrying all that is transmitted and any practical consideration thereof must regard the chromosomes simply as portions of the hereditary material. Whether one chromosome could by itself if necessary direct the development of an entire animal or whether the germs of different parts or organs are carried in separate chromosomes can hardly be conjectured. Regarding the chromatin simply as the hereditary material, with the facts that have been stated we can account for the lack of similarity in the offspring of two parents. Farmers who undertake to raise a pair of matched horses by breeding a mare to the same stallion in two successive seasons are frequently at a loss to account for the great disappointment.

Inheritance through Chromosomes.

Every cell in the body contains sixteen chromosomes, the direct product of the original group bequeathed equally by the two parents. In the preparation of the germ cells we know that one-half the chromatin bodies are eliminated. Considering at present, for the sake of clearness, a female of a species for which four is the regular number of chromosomes, we know that in a germ cell of that female only two chromosomes will be present to convey hereditary influences. It is impossible to know or

Possible Chromosome Combination in Ovum.

foretell which two of the original four bodies will be preserved. If we consider the four chromosomes to bear numbers from one to four, then while one and two may be in the ovum produced at one period, another ovum produced may contain the same combination again, or, it may contain numbers three and four or any two of the number present in whichever one of the primitive egg cells is being developed into an ovum. Any one of the following combinations is equally as likely as any other to be present in the ovum produced at any certain period:

    1 and 2,   2 and 3,
    1 and 3,   2 and 4,
    1 and 4,   3 and 4.

That the dam of an animal of a species of even four chromosomes should make exactly the same contribution to two successive offsprings is highly improbable, yet our larger animals have sixteen chromosomes to a cell.

The same considerations obtain on the sire's side. All divisions of the primitive spermatozoa remain functional, but only one is utilized in fertilization, therefore the probabilities are the same as with the female. Designating the chromosomes of the paternal cells as five, six, seven, and eight, a single spermatozoon has equal chances for carrying any one of the following:

Possible Chromosome Combinations in Spermatozoa.

    5 and 6,   6 and 7,
    5 and 7,   6 and 8,
    5 and 8,   7 and 8.

When fertilization occurs we know that some one of

Equally Probable Results of Fertilization. the equally probable maternal combinations will unite with some one of the equally probable paternal combinations. Any one of the pairs in the maternal list has equal probability of joining with any one of the paternal list.

Maternal	Paternal
1 and 2,	5 and 6,
1 and 3,	5 and 7,
1 and 4,	5 and 8,
2 and 3,	6 and 7,
2 and 4,	6 and 8,
3 and 4,	7 and 8.

The offspring will receive one of these equally probable sets of chromosomes:

1-2 and 5-6,	2-3 and 5-6,
1-2 and 5-7,	2-3 and 5-7,
1-2 and 5-8,	2-3 and 5-8,
1-2 and 6-7,	2-3 and 6-7,
1-2 and 6-8,	2-3 and 6-8,
1-2 and 7-8,	2-3 and 6-8,
1-3 and 5-6,	2-4 and 5-6,
1-3 and 5-7,	2-4 and 5-7,
1-3 and 5-8,	2-4 and 5-8,
1-3 and 6-7,	2-4 and 6-7,
1-3 and 6-8,	2-4 and 6-8,
1-3 and 7-8,	2-4 and 7-8,
1-4 and 5-6,	3-4 and 5-6,
1-4 and 5-7,	3-4 and 5-7,
1-4 and 5-8,	3-4 and 5-8,
1-4 and 6-7,	3-4 and 6-7,
1-4 and 6-8	3-4 and 6-8,
1-4 and 7-8,	3-4 and 7-8,

It is apparent that from two parents of a species with four chromosomes it is possible to have thirty-six individuals, no two of which would be identical. Of course, the majority would largely be of the same make-up, but in the two combinations written first and last there would be a very wide dissimilarity. The contrast of these two possibilities is the basis of the seeming impossible statement we sometimes hear, that two offspring of the same parents may be unrelated to each other.

If we consider our larger animals that are believed to possess sixteen chromosomes we find amazing possibilities. Taking the possible number of combinations of eight chromosomes that can be made up from sixteen in either parent and exhausting the number of unions that may be produced from these two sets, it is found that without duplication we may have combinations to the number of 65,536.

In view of the immense field of possibilities it is not surprising that we seldom find two animals that even seem to be identical or even nearly *Why Related* enough so to make a matched pair. *Animals Differ.* Some of the high-class animals produced by supposedly indifferent parents are doubtless the outcome of the rare occurrence of a combination of the best of material of each parent and the elimination of that tending to produce inferiority. Also some of the very mediocre offspring of renowned parents may be attributed to an exceedingly unfortuitous retention of the chromatin productive of inferior characters and the elimination of the desirable. It is undeniable that in this vital process of heredity there is and must ever be a large element of chance. Chance may govern what portions of the material will go to each offspring, but if it were possible to assure ourselves that all of each parent's supply was representative of good we might be careless of chance. However, it is not necessary or justifiable to assume that each chromosome is entirely different from all the others in the same or in another parent. In all probability they are largely similar. But they may be arranged in an infinite variety of ways and this arrangement is beyond all control. Though impossible to bring any influence to bear upon the manner of separating the portion of hereditary material for the new animal, we can yet assure ourselves of a desirable outcome by limiting ourselves to such animals as give us reason to believe that any selection from their stock of hereditary substance would contain the minimum of possibilities for undesirable characters.

Basis of Controlling Heredity. To achieve the greatest possible measure of control over heredity is the aim and need of the breeder of animals adapted to special uses. Heredity is chiefly if not entirely conveyed by the chromosomes of the germ cells. The elimination of some of these chromosomes and the amazing array of combinations that may be effected have all to do with determining the make-up of every creature. No degree of human influence over these processes is conceivable. How then has it been possible for the builders of breeds and types to mold the animal form so nearly to their liking? The answer is, by the selecting for mating of animals containing chromatin or hereditary material with the maximum possibilities of desirable features and the minimum of those undesirable; this done, no matter what hereditary bodies are eliminated or combined the result is still for good, and any few chance representations of unwelcome qualities are hopelessly in the minority.

CHAPTER VII.

ORIGIN OF THE HEREDITARY MATERIAL.

The phenomena discussed in the preceding pages give some conception of the fundamental nature of the forces to be dealt with. They suggest explanations of occurrences otherwise perplexing, but manifestly before we can undertake to formulate means to purify the hereditary material we must know something of its source. A consideration of the relation of that substance to the parent body is in order. The germ cells were traced from their rudimentary stages in the ovaries and testicles, but from whence did these organs derive this material of such extraordinary potency? We have arrived at the end of our positive knowledge of hereditary processes. No examination or experiment has as yet revealed all the facts regarding the immediate source of the contents of the sexual cells. Our embryologists explain development of the tissue layers and later of separate organs from the fertilized germ cell. But as to just what goes into the new reproductive organs there is no definite knowledge, though in the embryonic development of some lower forms there is evidence that an early segregation from the total parental germ plasm isolates a portion for re-

The Problem.

Necessity and Value of Theory.

serve in the new reproductive organs while the main amount is dissipated in the building of the body. In view of the great desirability of understanding the origin of this vital substance, the best that scientists can do is to theorize. That theory which accords with the facts and which best accounts for the various manifestations of heredity is the one which will be of greatest service.

Darwin believed that the reproductive organs acted in somewhat the same manner as do the secreting organs of the body, the material they secrete consisting of minute particles that enter the blood circulation from all the cells of the body and are withdrawn, in the ovaries of the female or testicles of the male, and built up into the ova or spermatozoa.

Darwin's Theory.

Weismann considers that the hereditary material is not drawn from the body but that rather a small proportion of the same material received from the parents is reserved intact in the reproductive organs and there remains until the animal reaches breeding age and then becomes active and produces the germ cells.

An Opposite View

These are the two main ideas on the subject. Actual examination or experiment to determine the facts seems impossible. We are therefore forced to base our practice on that explanation which seems most satisfactorily to explain the occurrences. If Darwin's suggestion be accepted we must chiefly emphasize the individuality of the parents rather than the ancestry, while the reverse is true if we think with Weismann.

In view of the fundamental importance of an intelligent idea of the source of the hereditary substance it is desirable and fair to more explicitly present the views of the two scientists referred to. Darwin's theory is known as "Pangenesis" and in his "The Variation of Plants and Animals under Domestication" he outlines his proposed explanation of the method of heredity in these words:

Pangenesis.

"It is universally admitted that the cells or units of the body increase by self-division or proliferation, retaining the same nature, and that they ultimately become converted into various tissues and substances of the body. But besides this means of increase I assume that the units throw off minute granules which are dispersed throughout the whole system; that these, when supplied with proper nutriment, multiply by self-division and are ultimately developed into units like those from which they were originally derived. These granules may be called 'gemmules'. They are collected from all parts of the system to constitute the sexual elements and their development in the next generation forms a new being; but they are likewise capable of transmission in a dormant state to future generations and may then be developed. Gemmules are supposed to be thrown off by every unit, not only during the adult state, but during each stage of development of every organism; but not necessarily during the continued existence of the same unit. Lastly, I assume that the gemmules in their dormant state have a mutual affinity for each other, leading to their aggregation into buds, or into the sexual elements. Hence, it is not the reproductive organs, or buds, which generate a new organism, but the units of which each individual is composed. These assumptions constitute the provisional hypothesis which I have called 'Pangenesis.'" He later states: "I am aware that my view is merely a provisional hypothesis or speculation; but, until a better one be advanced, it will serve to bring together a multitude of facts which are at present left disconnected by any efficient cause."*

Weismann's hypothesis is in the main the exact opposite of that of Darwin. He designates the chromatin or hereditary material as "germ plasm." His idea of "Continuity of Germ Plasm" regards the hereditary material as passing from generation to generation with the minimum of influence from, or association with the bodies of the parents.† He regards the ovaries and testicles as depositories of hereditary material. In them is deposited at an early stage of embryonic life, a part of the germ plasm, there to be retained intact until its host arrives at the age for reproduction. Before reproduction is rendered possible this dormant material within the organs quickens into activity; it absorbs food material from the circulation of its host, increases its volume and completes the various processes already explained as essential to the production of ova or spermatozoa. While at first thought it may appear strange to think of this germ plasm as living unmodified in the body from which it derives its support for increase, yet it is no more strange than the fact that widely different classes of plants draw from a particular soil those elements they need, and by virtue of its inherent tendencies each constructs the material into a new plant strictly of the ancestral type and but very slightly modified, if at

*Chapter 27, "Animals and Plants under Domestication."
†"Continuity of Germ Plasm" is fully presented by Weismann in a chapter on "The Germ Plasm."

all, by the medium in which it has its root. Even so may the germ plasm in the reproductive organs increase in quantity without changing materially its quality.

In the manner of its behavior subsequent to fertilization, in diffusing throughout the embryo and dominating every cell, the germ plasm is quite comparable to the yeast plant. What corresponds to the yeast supply is within the reproductive organs and is there perpetuated much like a parasitic growth, and periodically sends off portions of itself to grow and diffuse through a whole new organism just as the small portion of yeast multiplies, acts upon the flour in each batch of dough and changes it to a quite different product, while the yeast supply is continued indefinitely by affording the favorable conditions to the smallest amount of the original stock. The buttermaker carrying a good starter for a long period affords another analogy.

The Explanations Compared.

The many changes which animals undergo in the course of time would be accounted for by Weismann on the basis of selection from those departures or innovations occasioned by the necessity of reproduction by sexes, which process we have studied under the heads of maturation and fertilization. These vital considerations associated with the preparation and union of sexual cells as set forth in chapter five must not be confused with theory; they are fully demonstrated facts. It will be recalled that it was Weismann who saw the necessity for reducing division and predicted such a discovery before it was actually made.*

---

*Weismann, "The Germ Plasm," chapter 8.

The Darwinian theory would regard the germ cell as the epitome or concentration of the parent. While this might be conceivable for the production of most of the organs, a difficulty is encountered in the common case of persistence of lambs' tails in flocks in which sheep have been docked in early life for scores of generations. Darwin was not unmindful of this difficulty and met it by supposing the transmission of dormant gemmules carried down from early ancestors not docked, in sufficient numbers to reproduce such parts, of which the absence in parents would otherwise preclude their presence in the offspring because of the impossibility of such gemmules being present. In this case and in many similar ones, the influence of such dormant gemmules preserved from remote parents seems to be the rule rather than the exception. Weismann would regard the continuation of such characters not used or not present in the parents as regular and to be expected. Indeed his idea of continuity readily explains the persistence of such apparently needless parts as the vermiform appendage and the chestnuts on the legs of horses. In earlier forms these structures were doubtless functional. Under the Darwinian idea their disappearance and recurrence in offspring could only occur through dormant gemmules, as the exception; under the Weismann theory such occurrences would be the rule rather than the exception, as we know they are in nature.

If, however, we incline strongly to the theory of continuity of germ plasm we are apparently cut off from all possibility of the reflection in the offspring of even extreme conditions affecting their parents. It *is* admitted, however, that in extreme cases the lack or abundance of a specific substance in the system of the parent body may either retard or facilitate the multiplication within the primary cell of some part of the germ plasm dependent upon such specific substance. Of course, in case of under-support of such hereditary elements they would not be entirely excluded from the germ cells, but the same conditions obtaining for several successive generations would have a positive influence toward weakening such tendencies just as the opposite kind of support might strengthen them. This is the only provision made by Weismann for direct influence of environment upon heredity.* All else is due to the selection of the parents, governed either by natural demands or the artificial considerations obtaining in domestic animals. The selection of parents for their valuable qualities constitutes continual opportunity for modifying the make-up of germ plasm of the succeeding generations.

Relation of Practice to the Theories. Breeders who adhere to the idea contained in pangenesis would naturally judge of an animal's value as a breeder altogether from the characters he individually exhibits. His ancestry would be of interest only for chance of conveyance of dormant gemmules which would not be of more than very secondary importance. Although some breeders often permit what they deem a good line of ancestry to outweigh the individual characters of an animal, yet we must recognize the fact that all of our experienced and more suc-

---
*Weismann's views on this point are contained in a separate publication: "Germinal Selection as a Source of Definite Variation."

cessful breeders have been close students of pedigrees and their estimation of an animal's breeding powers has been based in large part on a knowledge of his ancestors. In so doing they have shown an appreciation of the chief principle of the Weismann idea, namely, that to mold our animals we must rely on changing the germ plasm by infusions by mating rather than by seeking to modify that substance by the influence of external conditions or being guided solely by external appearances. A conception of the nature of the germ plasm contained in an individual must be based upon a knowledge of the ancestors from whom that germ plasm was obtained no less than upon individual appearances. The practical significance of this principle is the subject of the next chapter.

CHAPTER VIII.

BREEDING AND SELECTION.

To say that the breeding of stock is fundamentally and chiefly a matter of selection is to repeat a truism.

Basis of Control of Heredity.

The primitive germ cells have been seen to go through important changes that determine what part of their contents shall be reconveyed to the next generation. Any tendencies or characters not represented in the material contained in the mature ovum and spermatozoon that unite in fertilization can not by any possible means be found in the new animal resulting from that union. Our only opportunity of controlling the make-up of the parental contributions to the offspring lies in becoming assured that in the whole hereditary substance of either parent there is nothing representative of objectionable characters; this being true, no matter what enters the embryo, the result is good.

Realizing that the parent does not draw from the various parts of its own body the components of the germ plasm we must not allow ourselves to regard any parent's germ cell as a recapitulation or even as a representation of itself. Viewing the ancestral source of the germ plasm and its comparative independence of the influence of the body, the off-

Why Offspring Resembles Parents.

spring then becomes an offshoot from the same stream that gave off the parent. Parent and offspring are similar because they have a common origin.

The numerous and distant sources from which any animal receives its inheritance are suggested in Fig. 7. The heavy lines from B and C, which enter A, represent the actual hereditary material contribution by those parents from the store in their own bodies, which was also implanted in each by their respective parents, the grandparents of A. A has not inherited and cannot transmit any tendency or quality that has not been contributed through his parents or grandparents. Of course it is possible and not improbable that a part of all of C's inheritance from G may happen to be represented in the polar bodies that perished when the ovum from which A developed underwent maturation, and thus A's inheritance through C may be stronger from F than from G or it may be the reverse or equal. This is indicated in the figure by the two lines of C's inheritance entering separately from F and G, the individuals contributing them, while the stream issuing from C draws from the combined supply an amount the same as entered from each of F and G, shown by the line leading from C being no larger than either of those coming from F and G.

A will be able to transmit good qualities in accordance with the degree of merit that was conveyed to him from his innumerable and distant

Ancestry and Prepotency.

progenitors through those that stand nearest him in descent. What possibilities were carried by these channels of inheritance in their devious windings and

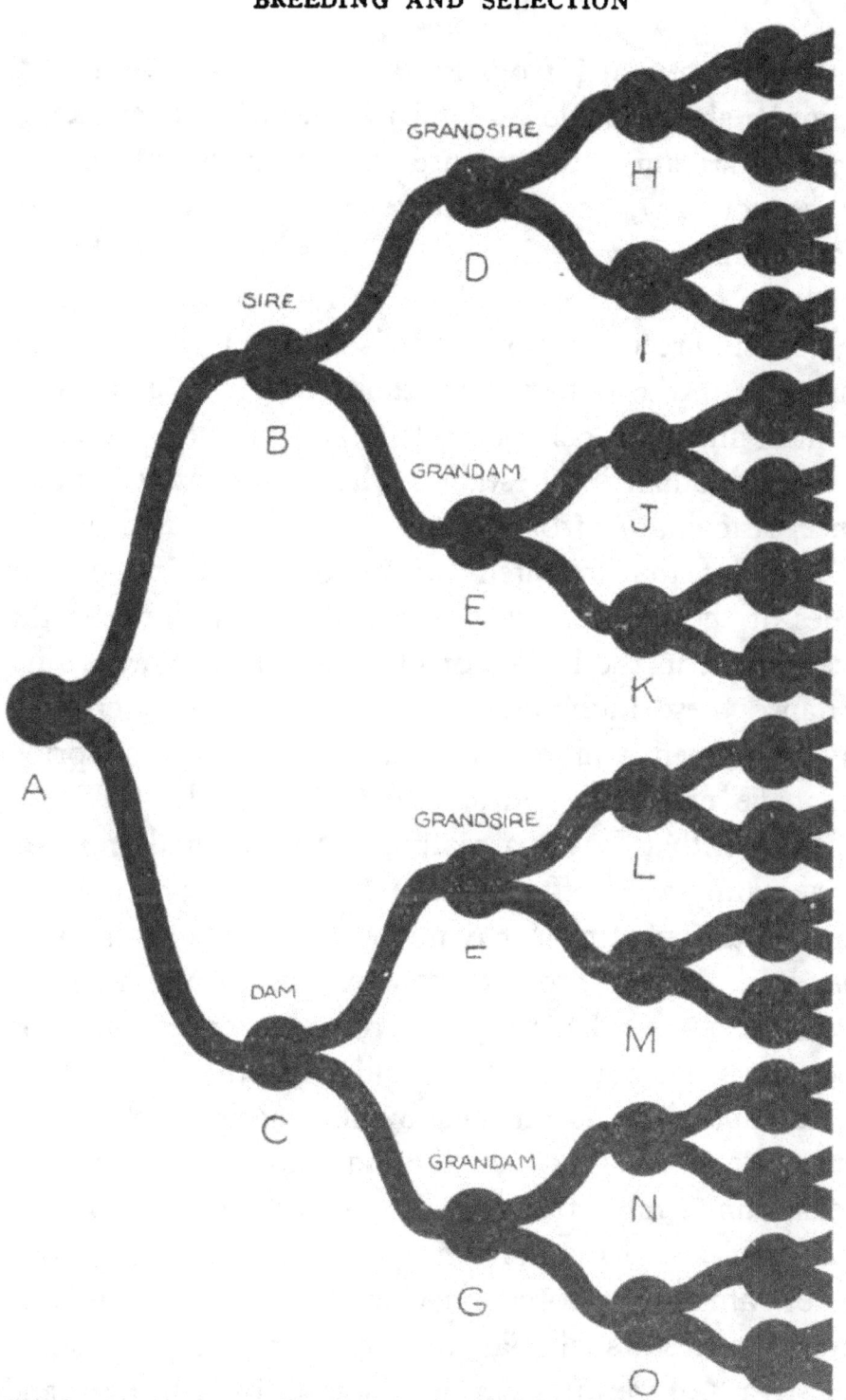

FIG. 7.—Represents the different lines or streams that enter into every animal's inheritance. The black circle in each case is the stored hereditary material within the body received from the lines entering on the right from the parents. The line leaving on the left represents the material of the germ cell given off to the individual it is shown to enter.

what was subtracted from the stores for each germ cell cannot be shown however far back we may undertake to trace the stream. To be sure the fundamental characters do not vary, but the features that give value to domestic animals are really minor ones so far as the resemblance of any individual to his race is concerned.

For the breeder's purpose it is sufficient to know the character of the material in those nearer courses that are most likely to contribute to what has been received by the individual in question. The nature of the inheritance, reaching A from E and J, can be shown by the development that resulted in the conformation of the bodies of those ancestors. It can also be judged by the development in the bodies of other animals to whose inheritance these individuals contributed. J's inheritance may have been a mixed one and some of his offspring may have exhibited undesirable features. If, however, it is known that E was a really good individual and produced mainly good offspring of which B was one, we may consider the stream as having been purified from inferiority in that part of its course. If examination of other lines shows that the flow from the sources of good inheritance has been added to only by other individuals whose superiority is attested by the merit of their offspring, we are assured that the individual in which these streams unite must transmit the excellence of his strain. This enables us to understand the strong breeding powers of animals whose inheritance traces exclusively through ancestors similar to each other in excellence. The stream of germ plasm has come to be of a pure and homogeneous makeup, and when mixed with that of an animal whose inheritance was not so restricted the pure material is able to dominate the miscellaneous tendencies from a mixed ancestry, and we have a pure-bred especially potent in stamping his likeness upon his offspring.

This figure also aids in explaining the phenomenon spoken of as atavism or reversion. The dam of H may have been of a red color, while the sire and all the other individuals to which B traces were black, and produced only black offspring. The same may be true on the maternal side except that N has an inheritance of red which has been present in both G and C, but held in check by stronger tendencies to black. If an ovum produced by C and containing a strong infusion from N of tendency toward the red color is fertilized by a spermatozoon from B that also happens to carry the remnants of a tendency toward red, then such a union may hold in check the tendencies toward the black color. It is unlikely that the same parents would produce similarly endowed germ cells at another mating, and thus their subsequent progeny would be of the usual color, as is commonly observed to be the case with parents of red Angus calves.

How Atavism May Occur.

To be assured of having a breeding animal that will transmit the maximum of the good with the minimum of bad it is then necessary to select one that individually exhibits such an inheritance, and that has had no ancestors from whom it might have received either active or dormant material to produce inferiority. As we proceed backward the probabilities of inheritance from any one ancestor diminish, but the possible preservation and recurrence of the contribution of that ancestor must always be reckoned with. The only practical method of directing heredity is to select for mating those animals that carry hereditary material of the desired potency and this can only be secured by the further selection of approved ancestors. This suggests an explanation of the preference some breeders of note have had for breeding the sires they used. In so doing it was possible for them to be more fully familiar with the ancestry and more competent to determine with what other descents matings should be made. The foregoing suggests as an ideal practice the selection of good individuals from good stock. These two factors, individuality and pedigree, are the subjects of the next chapters.

CHAPTER IX.

INDIVIDUAL EXCELLENCE IN BREEDING ANIMALS.

Foundation Stock.
Any measure of control over heredity attained by any breeder must be through the wisdom of his selection of the parents and ancestors of his stock. Care and feeding have their part and are indispensable as aids, but selection is the basis of the whole work. Whether the endeavor be to build up a herd or stud for the production of a uniformly superior class of animals for feeding for market, or to produce animals for others to breed from, selection is of fundamental importance. No haphazard unstudied procedure in selecting from the stock of others for a foundation, or indiscriminate culling of the increase of that foundation stock, can ever give satisfactory returns. It is not imperative that a person beginning the breeding of stock should anticipate and formulate a procedure for all possible contingencies, but the career of every breeder who has made himself known exhibits a quite clearly defined idea from the outset as to wherein his productions should differ from or accord with the various kinds and types to be found within his chosen breed. Later steps and plans may be decided upon in view of the outcome of earlier

work, but for best results there must be the recognition of a standard toward which to work.

It might easily be possible to acquire a large or small aggregation of foundation females with each one a superior individual in herself but unlike each of the others. Perhaps the diversity of types is nowhere more noticeable than in draft horses. At one time a ringside spectator may see the highest premium awarded to a very wide low short-legged squarely built horse while in the succeeding ring preference may be given to a horse of more lofty appearance, rounder, neater, smoother, a less massive but more active kind. Both types are useful, feed well and sell well. Some users find the horse of the first type well qualified to perform the work placed upon him, while in another line of business the second type is more serviceable. One judge with an inclination toward one type may send the premiums to that class of animals, while another judge at another time would give honors to the other. Complete agreement among authorities cannot be expected and is not desired. Both types are good property and fill their peculiar spheres of usefulness. The same holds true in other classes of stock. The type demanded by the cattle, hog, or sheep buyer is practically constant; but within each of the breeds of meat-producing stock there may be found types differing in size, rate of growth, rate of fattening, and grazing qualities, and consequently variously adapted to different sections of country or kinds of farming. The larger coarse later maturing and more rugged type may be more profitable to some men than the finer smaller

Types.

PERCHERON STALLION CHARACTER.

and more rapidly maturing kind, and each in turn may stand first in the showring and each may be valuable and salable.

Whatever may be true of the judge when acting officially, from the view point of the man who is rearing stock for sale the situation is different. It is to his financial advantage to have his young stock as nearly uniform as possible in type both for feeding and for selling. The same feature is of additional value to him who sells breeding stock. With a band of females of mixed type no one male could sire the same class of offspring from dams of varying stamp. There may be some very good ones from the inharmonious matings and some rare lucky results, but uniformity of appearance and strong power of transmission cannot result from such procedure. Nor could much uniformity be looked for in the increase if the sire were of one type and the dams were all of one but a different type. The selection of and adherence to a type is a more vital matter than the selection of a breed.

Value of Type.

It is the aim to insure the greatest possible amount of certainty regarding the outcome of every mating. Dealing with matters so imperfectly understood and so far from direct control it is not surprising that the unexpected should often happen. Because the unexpected may and does happen makes it all the more imperative that all possible means of insuring the desirable outcome be fully observed. The distinction among breeders lies not so much in the knowledge

Need of Full Study.

SHIRE STALLION CHARACTER.

they possess as in the thoroughness and persistence with which they utilize all available information. If a prospective member of a breeding herd be of the type with which it has been decided to work, it remains to make a careful study of its conformation and all its individual characteristics. The shape of every part of the body, and every feature, such as disposition and digestion, is inherited. We do not believe that an animal accumulates contributions from its various parts and organs to form the germ cells, but we do consider that the germ cells are a growth from an unused portion of the germ plasm in the fertilized cell in which that animal had its origin. Every desirable or undesirable feature about the animal is represented in the germ plasm, but unless every portion of that material be so strongly charged with the representation of any specific character as to insure its presence in every germ cell then that character is quite likely to fail of transmission. Undesirable features have exactly the same opportunity to be passed to the offspring as desirable ones, and the determination of what spermatozoon shall share in fertilization or what chromosomes shall be in the ovum is beyond direction. To preclude the possibilities of unwelcome characters in the offspring it is therefore essential that the parents' store of germ plasm be as nearly as possible free from possibilities for inferiority, and evidence as to this is to be had by making a thorough study of every part and feature. We have quite generally recognized and clearly defined ideas of good conformation in all types of animals. Fitness to wisely select breeding stock necessarily assumes a complete familiarity with at least the

SHROPSHIRE CHARACTER.

Breeder Must Be a Judge. class of stock in question and some experience in comparing and forming opinions of considerable numbers of individuals. Study of an animal's fitness to become a member of a breeding herd should extend farther than the visible or external features of conformation. In meat-making animals many of the required points of structure are merely indications of the capacity for consuming feed and producing maximum gains therefrom. Facts and records regarding the animal's feeding and producing qualities are usually obtainable and are many times more reliable as a basis of estimate than the most pleasing indications of the same capacities. With stock of which the usefulness may be made a matter of actual test, as is the case in dairy cattle, records of actual performance constitute the best possible evidence of individual merit, though age and various conditions affecting such a trial must be taken into account.

Prepotency. Among animals equally pleasing in build and apparently similar in efficiency of their special functions there often exists a marked variation in their power of transmitting their characteristics. Though such variation may frequently be due to differences in lineage yet certain features of individuality are found to be quite uniformly associated with power of transmission or prepotency. Prepotency in untested breeders is evidenced by that combination of physical attributes that gives to any animal a pronounced individualism, or as breeders term it, character. It is not easy to analyze character

HAMPSHIRE CHARACTER.   COTSWOLD CHARACTER.

into its component parts, but since it is so closely associated with prepotency an explanation of the term as used in stock-breeding is very desirable. Character, or the appearance of strong individualism, is contributed to by three things: style, high development of the appearances associated with sex, and that robustness and vigor of expression that can only be present where perfect health and spirits are coexistent.

Character.

Style as related to prepotency is allied more with breeding than with individuality. Its presence argues an inheritance from the animals produced by the foremost breeders who have always sought to combine attractiveness with utility. Appearances associated with sex, masculinity or femininity are often regarded as the main evidence of prepotency. We cannot recognize degrees in sex, but as in the case of a male the full development of the neck and front and the frontal bones of the face, though only secondary sexual qualities themselves, manifest the activity and full vigor of the functional organs with which they are connected. Likewise in the female the neatness of the neck and refinement of the features of the face, and the gentle disposition, all evidence the assertion of the female tendencies that have much to do with the young, both before and after birth. The robustness and vigor of expression read in the countenance and mainly in the eyes, and also reflected in boldness of movement, are probably the most directly associated with prepotency of all the things that may be regarded as contributing to character. The appearance and manifestation of maximum vigor and vitality can only be

INDIVIDUAL EXCELLENCE IN BREEDING ANIMALS 89

BERKSHIRE BOAR CHARACTER.

present where all organs of the body that have to do with digestion, circulation, respiration and the nervous system that controls all continuously perform their full work. This maximum efficiency of all organs makes up constitution and is indicated nowhere else so satisfactorily as in the expression of the countenance and in the general bearing, behavior and carriage.

The presence of this condition, the complete health and nourishment of the body, insures the highest vigor, vitality and activity of the germ cells. It cannot change the make-up of the germ plasm, but it may control its strength and power and thus give a much higher degree of prepotency than would be possible if the animal had been naturally weak or listless and low in physical vigor. The qualities that make up constitution and are therefore so closely akin to this character are inherent ones, represented in the germ plasm reserved in the parent for reproductive purposes, and therefore they may enter into the heredity of the offspring just the same as any other feature of the individual's physique. When possessed of such inheritance the offspring is imbued with the functional capacities that will enable it to withstand retarding and debilitating influences and, what is more important, to make the maximum response to careful and liberal feeding. The power to produce in proportion to the wisdom and liberality of the feeding is the fundamental distinction between improved and natural animals.

Significance of Character.

Sires of proved worth are often retained in active service until they reach an advanced age. So long as they remain in good physical condition and there is no noticeable decline in their impress upon their get there is no reason for regarding age as a factor in prepotency. When continued in service after the beginning of physical decline there is also a decline in the character of their progeny, showing again the relation to prepotency of an unimpaired individuality, and showing also the necessity of judging vigor by actual appearances rather than by the number of years through which the animal has passed.

Age and Prepotency.

Distinctive breed features such as color, shape of face or ear, or set of horn, are also a part of individuality as distinguished from pedigree. These features, commonly referred to as fancy points, while of no immediate usefulness are of considerable assistance in the selection of breeding stock. In the first place their presence is helpful because by the uninitiated they are regarded as trademarks, guaranteeing the presence of those special qualities on which rests the value and popularity of the particular breed they adorn. Where found apart from tangible evidences of actual utility they of course avail but little. It must be borne in mind, however, that it has ever been the object of the intelligent and far-seeing breeders to fix upon their stock such distinctive and attractive features as will commend them to the public and also appeal to and please the searchers after qualities of utility. In some instances selection has been based more on fancy than on utility points, to the great detriment of the latter, but when

Fancy Points.

found combined with proof or indications of real merit these fancy points serve as evidence of inheritance from the herds of the more discerning breeders and add assurance to the inheritance of and power to transmit the practical essentials.

According to the view of heredity pervading this discussion of individuality there is no possible means of a parent's transmitting to its progeny the effects of accidents or injuries. Exception must be made, however, to those abnormal conditions resulting from an inherited tendency toward such conditions. We do not believe that the germ cells carry representative material derived from each part of the parent body, but we do believe that the offspring will resemble the parent because they have a common source, and to be satisfied that the source is a good one we demand that the parent present high individual excellence as a proof thereof.

CHAPTER X.

PEDIGREES OF BREEDING ANIMALS.

Experience and science each afford abundant proof that rigidness of selection must apply no less to ancestors than to the present individual. We must judge of the hereditary material not alone by its accomplishment in a single instance but by its various sources and behavior in other instances of its existence.

Progeny the Best Test. It must be clearly recognized that as a basis of estimate of breeding powers nothing can compare with actual test, and where the progeny of a possible purchase are to be seen, individuality and pedigree both become at best secondary factors. In judging the results of a breeding test, however, it is necessary to have careful and full regard for the character of animals with which the individual was mated and the opportunity for development afforded the offspring. The parentage of increase of a fair degree of merit under limited opportunities is not satisfactory assurance of the ability to produce excellence when accorded the most favorable opportunity. It is but rarely, however, that an animal of proved excellence as a breeder is offered for sale, and selections have mainly to be made from untested stock on the basis of individuality and pedigree.

It is idle to discuss the relative importance of individuality and ancestry. One may be as valuable as the other in indicating what an animal will transmit; neither can safely be ignored or slighted and no breeder of note has ever failed to be a close student of both. "Individual excellence by inheritance" is the watchword of those whose stock gives them the most uniform excellence of increase.

Inasmuch as each parent contributed to the offspring equal amounts of hereditary material and also received equally from their parents in turn it is necessary to place equal emphasis on the paternal and maternal lines of descent. It is quite possible that the hereditary material bequeathed by one parent may be stronger for good or for bad than the contribution of the

Form of Pedigree

FIG. 8—TABULAR FORM OF PEDIGREE.

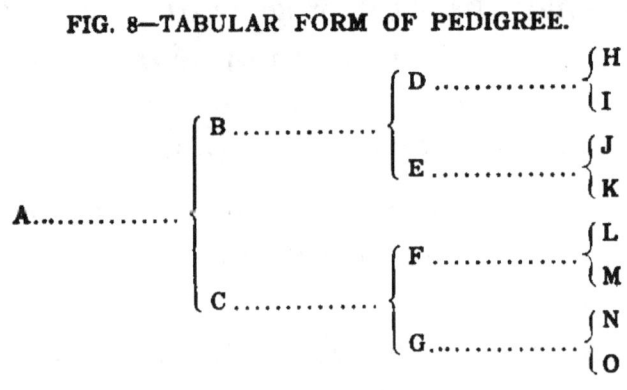

other. This may be due to more careful selection of that parent's ancestors, but it cannot be associated with either sex, and this further emphasizes the necessity of an examination of all the lines of descent. In arranging such lines of descent on paper for intensive study it is imperative that what is commonly known as the tabular arrangement shown in Fig. 8 be followed. Other forms of writing may be more economical of space and show a longer line of descent on one side, but for actual use in estimating breeding usefulness no other form is comparable with the tabulation which shows clearly every line of descent.

In studying an untested animal, represented by A in Fig. 8, whose individual make-up and qualities are approved, further evidence is needed *All Ancestors Must Be Studied.* regarding what may be contained in and transmitted by his hereditary material, because we know that any quality or character represented in A's germ plasm may appear in his get whether or not it was exhibited by himself. Crudely, A may be thought of as a composite or as an average of his ancestry, but from our knowledge of the facts of the preparation of the germ cells we recognize the possibility of having scanty or no inheritance from D, E, F, or G. It is also conceivable that there might have been handed down to him the impress of H or I or another in the same line much more strongly than from a nearer ancestor, all through the uncertainties of combinations of chromosomes or the seeming caprices possible in the formation of germ cells. It is therefore necessary to consider each ancestor as being represented in A unless tangible facts justify the conclusion that inheritance from any certain individual has been eliminated. Since it is manifestly impossible to understand the ultimate source of the germ plasm which A has inherited so variously, a study of what it has done in its more recent phases promises the greatest enlightenment in regard to its potentialities.

Since A is equally indebted to B and C we are naturally first concerned regarding those two animals and to them we may apply the tests we would prefer to apply to any animal in the following order: first, character of offspring; second, individuality; third, origin or breeding. The first named is usually practicable for parents and much greater value may be attached to the pedigree of an animal whose sire and dam are both proved to have produced offspring of merit. In considering the first produce of a sire or dam conservatism would at least suggest awaiting an opportunity to inspect subsequent progeny, which is usually no hindrance where sires are concerned. A few extra good and a large number of mediocre offspring would show the presence of potentialities for inferiority and compel the recognition of the possibility of a dormant inheritance of inferiority even in the more pleasing ones. Here, too, however, fair regard must be had for the opportunity for production of superior progeny afforded in their development and the choice of their other parent. Where the breeding test can be used it may properly outweigh all other considerations; in fact, some of the most noted matrons that have been frequent breeders and good mothers are far from attractive in appearance in their advanced years.

Breeding Records Of Parents.

When it seems desirable to still weigh the merits of an animal one or both of whose parents cannot be spoken for by their fruits, a full opportunity to study individuality cannot be foregone. In studying individualities of parents it becomes imperative to insist on their being at least very similar in type. Nor would even championship honors in close competition be sufficient assurance, because it is quite possible that under different judges or in different situations both male and female may have been accorded highest honors and yet represent types unsuited to each other. The progeny from such unions should certainly be required to first prove themselves capable of transmitting the blended excellence of their parents if indeed they have the unusual good fortune to exhibit a harmonious union of their divergent parental types.

Similarity of Type in Parents.

Showyard decisions at best constitute a very doubtful basis for the estimate of individual merit as a guide in breeding unless the selection is made by one sufficiently familiar with his work to be able to make necessary allowance for official opinions and subsequent changes of form. In most classes of stock the show records of the progeny of individuals in the pedigree under study will need to be relied upon to furnish evidence of their rank as breeders. A show record may do more or less than justice to a single animal, but applied to what his offspring have done in the ring it is almost sure to represent his actual standing in his breed. Due consideration must be had for probable variations in opinion of judges and for the inequalities of competition on different occasions and at different places. Then, too, in weighing the achievements of the progeny

Value of Show Awards.

of a particular sire or dam undue stress must not be laid upon a single offspring of phenomenal record to the exclusion of others of no note.

Inheritance from a sire most of whose get could earn even fourth or fifth position or even honorable mention in harder competition would be much *Fair Estimate* preferable to that from an animal *Of Sire.* siring one champion and no others of more than very local repute or of fame borrowed from their kindred. Also in many instances a second premium is practically as honorable as a first in spite of the fact that nearly all the general acclaim is accorded the holder of the end position. The prizes for get of sire and produce of dam awarded in our shows are the most valuable of all for showing the prepotency of parents.

Strange as it may seem, not one of the breeds has any official register of the results of showyard trials. One very laudable attempt was made by a Hereford breeder to establish a "star list" which was arranged to show with a minimum of searching the achievements of every winner and producer of winners in the larger shows. Such publications, to fully meet the wants, must be prepared by persons who cannot be thought of as having any interest in any animal, herd or strain. Some of the beef cattle herd books have appended lists of awards at leading shows, but so far these are not arranged to encourage even an anxious inquirer to attempt to procure the record of a particular animal. For the most part, information of this character must still be obtained from the history as recorded in the agricultural journals and periodicals and from association with persons in whose memories the facts have been preserved.

With dairy cattle and race horses the records are much more useful. While showing is popular, merit is proved chiefly by actual test of function. A record of having produced 20 pounds of butter in a week, or of having trotted a mile in 2:15, requires no consideration of errors in judgment or unworthy competition. The standard is an absolute one and can be applied at any time or place. It is possible that such records may not fully represent the capacities of the individuals because of limited opportunities, and especially with the cows a knowledge of food consumed during the test is desirable, but there can be no gainsaying that fact that under conditions surrounding the trial the animal possessed the ability to perform as recorded. Such trials also render it easy to state the achievements of the progeny of any sire or dam. The information made available in the "Year Book" for trotting horse breeders is of the greatest service in selection and study of ancestry and is doubtless in large measure accountable for the remarkable accomplishments in breeding for trotting speed. Tables similar to those in the "Year Book" may be forthcoming for dairy breeds as soon as official testings have been in use for a sufficient time. Although the significance of showring prizes is less dependable than test records it would seem that a great help would be afforded breeders of other classes of stock by preserving and publishing well arranged show records and compiling tables showing sires and dams with lists of names

Advanced Registers.

and achievements of their progeny that had been exhibited.

The question may properly be raised, is it safe or fair to withhold our esteem from progenitors, which though worthy, were allowed no opportunity to make a show career or to take a record? Doubtless some animals of extraordinary capacities have been allowed to live and die in comparative obscurity. Such individuals must necessarily have been the property of men not active nor prominent in the affairs of the breed handled; otherwise the merits of their stock would have been made known. If such an animal were unrightfully retained in obscurity with no opportunity to justify himself through his offspring, the probability of underestimating any valuable inheritance from him is very small because his excellence must have died with him.

Obscured Merit.

Animals without offspring to speak for them, besides standing on their individuality must also lean in turn upon their parents, and even when no such lack exists the grandparents must be well scrutinized to afford a fuller knowledge of the inheritance and possibilities that may have been imparted to the descendant. Grandsires and grandams must be measured by the same standards as were set up for the first parents, namely, character of offspring, individual merit, and ancestry. Here there will always be opportunity to learn what has been achieved under actual breeding test and this consideration will outweigh the other two. It is necessary, though, to be assured that grandsire or grandam, as the case may

Grandparents.

be, has transmitted the good features, and if undesirable ones do exist, that they have been counteracted in the selection of mates and are at least less prominent in the succeeding generation. The third section of our standard carries us into another generation and the question naturally arises as to how far we must carry this study. It is altogether reasonable to place greatest emphasis on the more recent progenitors and correspondingly less on those more remote. The study of pedigree is an effort to understand through an examination of its various sources the nature of the accumulated hereditary material. The further back we can trace the course of its flow and the more exhaustive our scrutiny of the various tributaries or sources of supply, the more dependable and complete is our information. A long line of ancestors with records of having produced the minimum of inferiority and of having continued to produce uniformly in accordance with their own type is the strongest possible and only conclusive proof that the hereditary material has been fully purged from all impurities by careful selection exercised by the breeders of those former generations in their elimination of all ancestors exhibiting or producing undesirable qualities.

Near and Remote Ancestors. A breeder is likely to meet with two other types of pedigrees, one in which the first three or four generations show animals of merit as individuals and as breeders but in which the back lines show few familiar names and represent obscurity if not inferiority. The other kind of breeding is more common, that in which the fourth and more remote lines show many animals of fairly earned distinction but in which the nearer generations have not come to fame and seem to rely more on their descent than on themselves or their performances. Either one of such pedigrees must be considered much less valuable than one in which all lines are of proved superiority, but of the two, the one with obscurity surrounding near relations is inferior to the one with distinction in close lines and obscurity in remote lines. The esteem in which this latter style of pedigree is sometimes held has prompted some well meaning writers to decry as a snare and a delusion the whole matter of pedigree. Certainly in its abuse it does present an insidious danger which has brought loss and disappointment to many. Z in the tabulation shown in Fig. 9 typifies the kind of breeding under discussion.

A may be taken to represent a sire of earned popularity and of whose sons A 4th proves to be able to beget stock of more than ordinary merit. This fact when properly advertised by his owner, creates a strong demand for his offspring. In the haste and eagerness to secure such stock, individual merit of the purchases is ignored, or else it is hoped that the offspring of A 13th will resemble A 4th rather than their own sire. At other times it is expected that the continued popularity of the strain will continue to unduly attach itself to the descendants and enable them to sell in spite of their defects. In the desire to profit by the popularity of A 4th his owner may mate him to inferior females and retain such offspring, of which A 13th may be one, that show plainly that they inherit more deeply of the defects than of the excellencies of their sire, or they may show that the female Q was not well adapted to mating with A 4th. The same blind adherence to U, an inferior descendant of the really meritorious B 3rd, gives X a double infusion of the inheritance from what should have been the rejected offspring and misrepresentatives of really good breeding individuals. Confidence placed in Z solely because of his

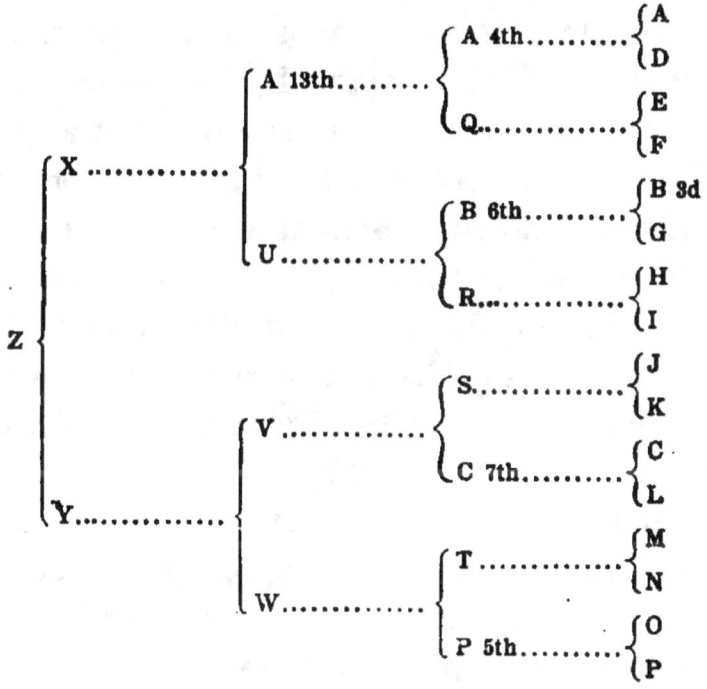

FIG. 9—TABULATION SHOWING RELATION TO A DISTINGUISHED ANCESTOR.

kinship to A and B 3rd ignores the fact that less than one-quarter of his inheritance comes from these two animals while the remainder is from others inferior by inheritance. While it must be admitted that such as Z will often find buyers, and it might be possible to justify traffic in the kind because others erroneously overrate them, yet the probabilities of his transmitting the characteristics of his few distant ancestors of note are too trivial to justify regarding his use as other than a random experiment. Real or pretended faith in the value of such a pedigree represents the abuse rather than the use of a study of the ancestry and is damaging because it ignores the absolute necessity of first applying the tests to the members of the nearer generations. The appearance of the name of the most distinguished animal in the fourth or fifth line signifies very little. Earned popularity as a sire attaches to sons and grandsons, the inferior as well as the better ones being sought for by less discriminating buyers. All the sires in use in any breed at a given time trace to a surprisingly small number of predecessors. Consequently almost any animal will trace once to a well known individual, and if the attention is allowed to centre mainly on the remote lines practically all animals will be found able to boast of distinguished ancestors in common. "A worthy son of a worthy sire" expresses the principle that cannot safely be lost sight of.

Unusual performance under test or extraordinary showring success often causes a very eager demand for the offspring of the animal so elevated, or for others so nearly related as to have promise of producing similar excellence. Such a strain or family then becomes fashionable and rightly so, because the fashion proceeds from incontrovertible merit. It is only when there is an indiscriminate acceptance of unworthy representatives of worthy families that the fashion becomes a blind craze, with the deteriorating influences referred to in the preceding paragraphs. The reference to fashion at this time, however, is made for the purpose of introducing the matter of family names. It is the custom, particularly in some breeds of cattle, to lay stress on family names. It is argued that among so many herds and varied strains of breeding there is need of such names as shall furnish some information regarding the line of descent. In the human family names are usually preserved through the male line. A person bearing the name of Smith may have scores of ancestors of other names and nationalities for one of the Smith family, and it is but rarely that different families of the same name have much more than the name in common unless they are otherwise akin. Nevertheless, the use of the name is a necessity whether or not it is any suggestion of family characteristics. The claim is made that a similar system is desirable for use among animals and since the name of the sire attaches to so many individuals the name of the dam is used instead. The right of use of any particular name is accorded only to those whose ancestry traces exclusively through females to the foundress of the family.

Fashion and Family Names.

It seems likely that family names came into use more through incidental causes than as a designed compliance with an actual need. When the breeds were being formed and when a very few herds included all the better stock some females were especial favorites with their owners because of their excellence as breeders. It was much more definite to refer to a calf as a son or grandson of the cow Duchess than to designate him as the offspring of a sire whose get included a large number of individuals of various maternal ancestries. Certain females transmitted particular and valuable qualities and it naturally became advantageous to own animals closely related to such foundresses of families. While a son might be equally as valuable as a daughter for perpetuating those qualities it was to the advantage of the owners to apply the family name only to descendants through the female line. It was thereby practicable for them to retain in their own herds as many as they chose of such descendants, while the males leaving the herd would share the prestige of the family but could not add to the numbers of those entitled to the family name. At the time referred to herd books were not established and no printed pedigrees were available, so statement of membership in a particular family was useful even if only partial information regarding breeding.

In America it is customary to recognize imported females as originators of family names. So long as the descendants exhibit the characteristics that popularized their family name they are rightfully entitled to any preference attaching thereto, but when it amounts to the blind or unintelligent scramble for the discards of those families and becomes purely a matter of name, only injury can result. The animal Z in the tabulation of page 103 is a member of the P family, even should another female appear a half dozen times in the same line, because the female descent is unbroken only to P. On the other hand, light esteem or prejudice is sometimes attached to descendants of females blacklisted by owners of contemporaneous stock, or by an unfounded suspicion, in spite of the fact that the animal regarding which the question is raised cannot at most derive 1 per cent of its inheritance from the defamed progenitor. That family names are not a real necessity is made clear by the continued advance made by most of the breeds in which no preference attaches to direct descent from one matron over that accorded to the same possibility of influence through intervening male ancestors. In such breeds as retain the custom it is not the rule for the animal's recorded name to contain any part of the family name and it seems entirely probable that use of family names will soon be altogether abandoned. With officially recorded names and numbers for each animal and easily obtained complete pedigrees the need of indicating descent in a name no longer exists, though it is a useful practice to give immediate offspring of a well known male or female such names as will suggest their parentage.

The matter of judging pedigree, like that of judging the animals themselves, is much more simple in theory than in practice. Even were it possible to obtain all desired information regarding a pedigree, there is no possible form of expressing in abstract terms the measure of its value, but to one who has a wide and impartial knowledge of recent and current happenings it is quite an easy matter to arrive at a safe opinion of the total value of the ancestry of any animal as presented in a well written pedigree. But it is in securing such information as is sure to be desired that one of the practical difficulties arises. Applying the triple test of character of progeny, individual merit, and breeding, to each ancestor appearing in the tabulated form, it may often happen that some ancestor near

Significance of Breeders' Names.

enough to be of importance will be unknown except for the name of its breeder. Impressions of outstanding individuals, and of many less notable but more familiar, are easily retained. A practical way of supplementing such knowledge is by studying the breeders.

Any breeding enterprise, sooner or later, will have a rating in public esteem in accordance with the soundness of the principles actually adhered to by the breeder. One may know nothing of a particular animal, but if he learns that it was reared by a man who is known as having always exercised the most careful discrimination in the selection of sires and the culling of females he will be assured that there is at least a preponderance of inheritance for good. If it is known that the third, fourth, and successive dams were bred by the same breeder whose achievements had brought him the esteem of his contemporaries and who would not retain an inferior female in his herd, then the standing of the herd attaches to its descendant. In such a case the breeder's methods are a guarantee that none but good sires were used, but if in addition we learn that those sires were from other herds of the best repute then there is good reason for placing a high valuation on an animal of such descent even though the particulars regarding the ancestors are very meagre. The custom of naming animals so as to include the name of the breeder or of his farm is a very great help in this connection, though it must be observed that it is quite common to continue to give such names to the descendants of such animals bred by other parties who do not exercise equally careful selection. It is not impossible that a third or fourth sire or dam valued for reasons just discussed was discarded for failure to represent the type and features sought for by the breeder. If direct evidence as to individual merit of such is not at hand it will be necessary again to place dependence upon the standing of the breeder who owned the animal at the time the offspring concerned was bred. In the absence of direct information the most conservative procedure will make the standing of the breeders the main part of the basis of opinion of the value of more distant ancestors.

While it may not be easy to gain full acquaintance with the past, the study of current events in shows and sales is a very interesting and profitable investment of time. Generations of animals come and go very quickly and a man conversant with one or two seasons' affairs soon finds the subjects of his study appearing in the fourth and fifth lines of pedigrees and his knowledge ample for the nearer and more important ancestors. The association with men whose knowledge antedates one's own is a most useful means of studying breed history.

In a few rare instances breeders have been known to represent an animal as being the offspring of a parent much superior to the actual one. With the magnitude and character of the business no possible means can be employed to verify the representations of breeders in these matters. No more reprehensible form of dishonesty can be conceived than that which would cause a breeder to stake his judgment and the value of even a single crop of young stock upon an animal whose descent is not as represented. Careless-

Correctness of Pedigrees.

ness in keeping of records may lead to unintentional errors, but the proportion of thoroughly careful and reliable breeders is so great that there is no necessity for dealing with any party who allows any question to exist regarding the honesty or correctness of his representations. The following pedigree score card suggests the relative importance of near and remote ancestors and the basis of estimating their influence:

```
┌ Record for uni-      ┌ Record as a sire.4  ┌ Record as a sire.....1
│   formly   siring   │ Individuality ....2 │ Individuality ........1
│   good stock....12  │                      └ Ancestry ............1
│ Individuality ...12 │                      ┌ Record as a producer.1
│                     │                      │ Individuality ........1
│                     │                      └ Ancestry ............1
│                     │ Record as a pro-    ┌ Record as a sire.....1
│                     │   ducer of good     │ Individuality ........1
│                     │   stock ........3   └ Ancestry ............1
│                     └ Individuality ...3  ┌ Record as a producer.1
│                                            │ Individuality ........1
│                                            └ Ancestry ............1
│
│                     ┌ Record as a sire.4  ┌ Record as a sire.....1
│                     │ Individuality ....2 │ Individuality ........1
│                     │                      └ Ancestry ............1
│ Record as a pro-    │                      ┌ Record as a producer.1
│   ducer of good     │                      │ Individuality ........1
│   stock ........10  │                      └ Ancestry ............1
│ Individuality ...14 │ Record as a pro-    ┌ Record as a sire.....1
│                     │   ducer of good     │ Individuality ........1
│                     │   stock ........3   └ Ancestry ............1
│                     └ Individuality ...3  ┌ Record as a producer.1
│                                            │ Individuality ........1
│                                            └ Ancestry ............1
│ Similarity in type
└   of sire and dam 4
        ─────            ─────                    ─────
         52        +      24        +              24 = 100
```

CHAPTER XI.

THE OFFSPRING DURING GESTATION.

The successive stages of development from a fertilized ovum to a fully formed body occur with striking regularity and uniformity. These changes, however, are of more immediate interest to the embryologist than to the practical student of heredity. But even the latter should not lose sight of the significance of one of the stages which was referred to in Chapter VI, the separation of an amount of the germ plasm or hereditary material to be preserved in the ovaries or testicles as the case may be, for the production of the germ cells of the succeeding generation. This idea renders more easy the appreciation of the importance of good ancestry.

It is known that in most animals some days elapse before the fertilized ovum becomes attached to the uterus to be sustained from the blood circulation of the dam. During the interval the changes that occur are supported by the considerable amount of food material carried by the ovum or egg-cell. It was noticed in Chapter IV that there is a natural tendency to suppose that the very intimate contact existing between the foetus and the dam through such a long period of time affords her extraordinary opportunity to

Relation of Foetus to Dam.

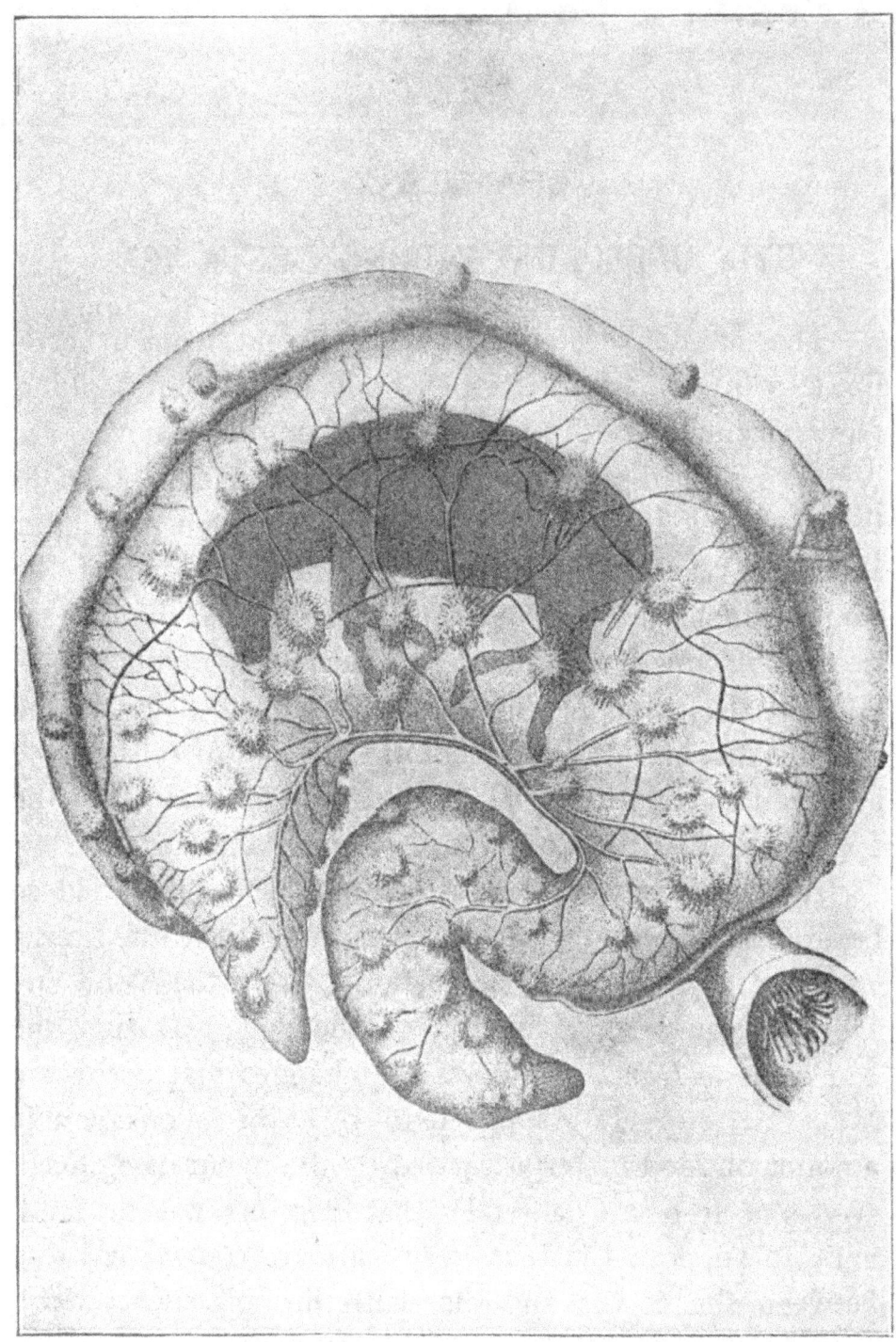

FIG. 10.—FOETAL CALF WITHIN ITS MEMBRANES.—Laws, "Diseases of Cattle," Bureau of Animal Industry, United States Department of Agriculture.

THE OFFSPRING DURING GESTATION 113

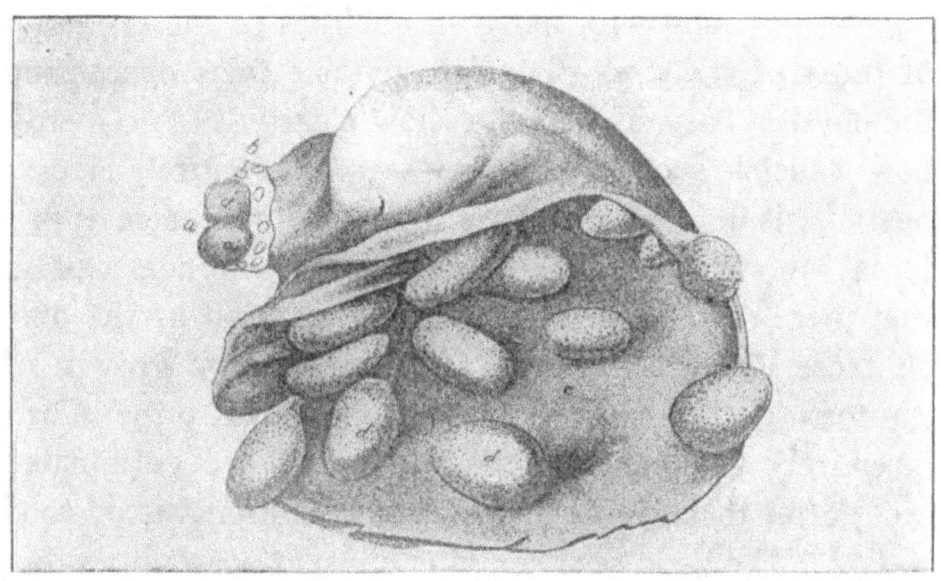

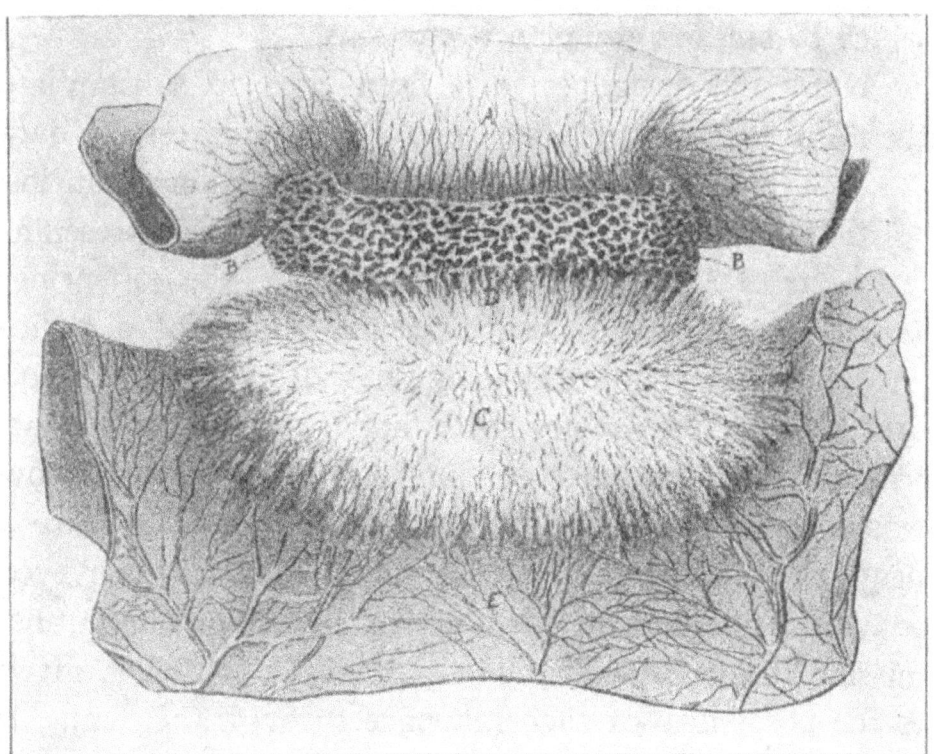

FIG. 11.—PREGNANT UTERUS WITH COTYLEDONS, AND A MATERNAL COTYLEDON, B. B., ATTACHING THE ENVELOPING MEMBRANE OF FETUS, E, TO UTERUS, A.—Laws, "Diseases of Cattle," Bureau of Animal Industry, United States Department of Agriculture.

imbue the young with her own qualities to the exclusion of those of the sire. The indisputable facts concerning the physical basis of heredity show clearly that no matter how plausible such an idea may seem it is entirely erroneous. This in no way detracts from the importance of maternal vigor and good care and feeding to render certain that there shall be present in the circulation of the dam all those elements requisite to the maximum growth of the foetus. The attachment of the embryo to the uterus is entirely analogous to the rooting of the germinated seed; after that point everything is conditional upon food supply. The good seed contains unusual possibilities, but is helpless and useless in the absence of the material with which to build to the plan it contains.

Numerous instances have been cited to substantiate the claim that certain conditions may so impress the dam at the time of conception or during pregnancy as to cause representative conditions in the offspring.

Effect Upon Foetus of Maternal Impressions.

The writer has known of a Galloway cow that dropped an off-colored calf. The owner with apparent seriousness attributed the occurrence to the fact that before the cow was bred the herdsman had allowed the bull to serve a neighbor's family cow the color of which was conveyed to the calf in question. It is unusual to claim such transmission through the sire, but credence is too often given to the possibility of such influence through the dam. It is related that the celebrated Angus breeder, McCombie, attributed some of his success in ridding his herd of the tendency to throw white spots and off-colors to his having painted all his barns and fences in a solid black color so as to impress his breeding animals.

That impressions upon the mind of the pregnant mother are reflected in the offspring is completely impossible of explanation on a physiological basis. It is true we are very unfamiliar with the nervous system, but to suppose that even violent mental impressions could originate a substance that would so derange inheritance as to produce a serious change in one organ or part while others are not affected would be going very far to explain even undisputed facts. It is also well to bear in mind that the circulation of the dam does not pass directly through the foetus, but that the latter has its own system of circulation which is replenished by filtration through the numerous cotyledons that connect the inner maternal and the outer foetal membranes.

That a mental impression could set up an action that would be conveyed to a specific part of the foetus is unthinkable. There doubtless have been cases where animals were born with some deformity or malformation corresponding to a condition that impressed the dam during pregnancy. But such cases are so very rare as to compel us to class them as coincidences. For every such case that can be cited, there are thousands of others in which the same or equally likely influences were exerted with no result. If it were true that visual impressions could be conveyed to the offspring, breeding would be chaos. Colts would have the color of the cattle or swine, the calves conceived in summer would be of a green color and those of winter would reflect the varied hues of the surroundings of the yards and the interior of the stables. Speculation and discussion in regard to these very rare coincidences diverts attention from the tangible basis of heredity which alone yields suggestions that can safely be carried into practice.

The embryo must not, however, be regarded as unsusceptible to the effects of the mother's condition. Although maternal impressions cannot be directly conveyed to a specific part or organ of the foetus, regard must be had for the fact that severe nervous disturbance occasioned by fright or anger may interfere with nutrition. Authenticated cases give best of grounds for believing that anger may so derange the nerves and the organs they control as to cause an abnormal and injurious condition of the milk. Since milk is a blood product it is reasonable to suppose that the same malnutrition may also extend to a foetus in the uterus and cause a partial or complete interruption of nutrition of the foetus and death or expulsion or both. Such possibilities suggest the general precautions against allowing infoal mares to be in sight of blood, and against the feeding of damaged feed to any stock carrying young. Strange and mysterious marks and conditions may also be the result of the displacement of the foetus and pressure or entanglement of parts in the cords in such a way as to cut off the circulation to the part. Such conditions suggest the protection of the dams from undue exertion and rough treatment which may also cause abortion or the death of the young.

Need of Care in Management.

While it is true that no regard need be had for direct influence of maternal impressions and that reasonable treatment will preclude accidental abortion, still the time of gestation affords important opportunities to second the efforts put forth in selection. Reproduction is a normal function, and only normal treatment of the parents is necessary to insure its successful accomplishment. While as with plants improper or insufficient nutrition may not produce specific effects, at the same time the foetus can complete no development for which the material to build with is not forthcoming from the dam's circulation. The hereditary material represents the ability and powers of architects, but the most expert architect is just as helpless without necessary materials as is the man surrounded by material but without masons or carpenters.

Nutrition of Offspring.

The securing of maximum development before birth has a very important relation to the outcome of any mating. The necessity of liberal feeding of the mother to insure a plentiful supply of milk is easily recognized, but the beginning of suckling, while it is a vital transition to the offspring, for the dam marks only a changed method of nursing. The nourishment of the offspring prior to birth may have just as strong an influence upon its final development as that furnished after it enters upon a separate existence. Any meagerness of the feeding during pre-natal days impairs and restricts the development of all the organs. Under favorable circumstances such under-development may be overcome by careful feeding after birth, but such procedure consumes time that might have been utilized in making progress toward maturity and never can fully compensate for curtailment at the more opportune time. Though prematurely born animals that are exceptionally well tended sometimes mature well, cases are not infrequently met with in which lack of vitality and many serious forms of weakness are traceable to under-development at birth.

The support of the growth of the foetus through the feeding of the dam must be considered as being accomplished only after the demands for her own sustenance and whatever may be exacted in the way of milk or labor have been satisfied. Nor is a fat condition of a pregnant female evidence that the feeding is judicious. The fat producing feeds are not what is chiefly required by the growing young. The increase consists mainly of bone, muscle and body tissue, and must be furnished in the dam's ration. Liberal supplies of unwisely selected feeds are not of themselves a guarantee of the most desirable results. Unhealthy stabling, poor ventilation or restricted exercise may preclude the most healthy and efficient condition of the dam and thus hinder the young from accomplishing what was made possible in planning its inheritance.

Feeding the Dam.

The bovine embryo at the end of the third month has a length of about 5½ inches. At the end of the fourth month the length is in the vicinity of 10 inches and the weight 4½ pounds. Up to this point the tax upon the dam has not been severe and might be supported while she devoted considerable of the food products to labor, milk and growth. During the fifth month length and weight both increase

Growth of Bovine Foetus.

by about 50 per cent. It can readily be seen, then, that in the case of a calf weighing 80 pounds at birth there is an increase in weight of over 70 pounds during the last four months. This weight consists nearly altogether of bone, muscle, and other tissue. It represents a gain through growth of over half a pound daily, that must be supported from the dam's feed. If her ration is lacking in the required elements the young must most certainly suffer. If she is so fed and handled as to continue a heavy milk flow that demand is not unlikely to be supplied at the expense of the fœtus. If her growth is incomplete, and the ration not liberal enough to meet the needs of two individuals, one and probably both will suffer. One very successful breeder and exhibitor with whom the writer was acquainted stated that his best calves in a surprisingly large proportion of cases were from cows that had missed breeding the previous year. This breeder considered that the rest permitted the cow to be in the best possible condition to nourish her young both before and after its birth. Reasonable exercise at liberty or at work may promote the growth of a foal through the general health of the mare, but severe labor or even modest labor, when only fat producing foods are fed, can be exacted only by sacrificing in some measure the natural vigor needed by a well bred animal.

Influence of a Previous Impregnation. Before proceeding to a consideration of the interests of the offspring one other topic may be treated. Along with wonderful tales of the effects of maternal impressions there are recorded instances which are taken to show the residual influence of a sire upon other later offspring produced by the female to the service of a different male. This is designated as telegony, or the influence of a previous impregnation. Belief in such a supposed phenomenon is illustrated in the idea that a mare raising a mule colt, though mated with a male of her own kind the following year, will produce a foal exhibiting characteristics of the ass. The majority of breeders have no regard whatever for any possible influence of earlier sires because their experience and observations do not so suggest. It used to be claimed that telegony was operative in dog breeding, but the number of dog breeders who believe in it is rapidly decreasing. Supposed occurrences in this class of stock can be explained on other grounds. There is no satisfactory physiological explanation of telegony. The facts do not suggest that such a thing exists; it merits no practical or speculative consideration by breeders.

CHAPTER XII.

DEVELOPMENT OF YOUNG STOCK.

An animal's inheritance is complete at the instant of conception. Everything he is to be by virtue of his parentage and ancestry is already implanted. His food, protection or lack of it, training, and everything connected with his subsequent life make up what is spoken of as environment. Many warm and earnest debates have been occasioned by differences in opinion regarding the relative importance of heredity and environment and the transmission of the effects of environment.

What Constitutes Environment.

The stock-raiser's interest in environment is in two phases: first, its relation to the individual animal, and second, its effect on the offspring of that individual. The second concerns only the raiser of stock to be used for breeding purposes, while the possible influence of environment upon the individual is of immediate interest to every stock owner.

The practical relation of environment to heredity in the development of the individuals will be considered first. It is a common remark that the influence of environment upon farm animals is a deteriorating one; that environment is more powerful than heredity because when extra care and feed are withheld from the young the development that characterized the parents is not secured. In the same line of reasoning heredity is held to be secondary to environment because seeming low-bred animals reach unusual development under favorable opportunities, and the "corn crib cross" is advised as a chief factor in improvement.

No good can come from debating the values of heredity and environment, since each is essential, and any lack in one will curtail the possibilities of the other. Inheritance is mainly a matter of possibilities. The advantage of the good pure-bred over the scrub is chiefly in his greater possibilities. Of course he exhibits the color and external features of his breed, and the peculiar conformation, but any advantage through ability to derive a greater amount of nutriment from a given amount of feed is at best very slight. The chief distinction lies in the fact that the pure-bred, under the direction of his inherent nature, constructs from his feed a body of greater value, and in the additional fact that he can consume a greater amount of feed. The power of greater consumption is a decided advantage even though the degree of efficiency of digestion be the same as in the scrub. A very considerable part of what an animal can consume is required for maintenance and only that amount of food digested in excess of needs of maintenance can be used for gain. Consequently the one that consumes the most can devote a larger total amount to purposes of increase, complete his growth more quickly and effect an economy equal to the cost of

Improved Stock for Improved Environment.

maintenance during the extra time required by the scrub. Under a system of low feeding and poor care, the natural environment under which the scrub's ancestors were reared, the improved animal has no opportunity to utilize his inheritance and makes an indifferent showing against his rival that is in his own habitat. Improved environment is an imperative adjunct to improved breeding. Under judicious artificial care and feeding, the artificial environment that surrounded the ancestors of the good pure-bred, opportunity is offered for utilizing inherent possibilities and the result is markedly in favor of the improved individual.

The scrub is the outcome of natural selection to answer requirements of natural environment, and if stock is desired for the purpose of withstanding the adversities of poor feeding and treatment, the scrub will admirably fill the bill. Our breeds of improved stock have been evolved by artificial selection to meet the needs of the artificial methods of rearing and *Feeding Must* use obtaining in all advanced agri-*Support Breeding.* cultural sections. Their superiority cannot assert itself in the absence of the accustomed environment, and when we assume the presence of natural conditions or a low order of care then environment does assuredly tend to pull down or hinder the assertion of what artificial selection has built into heredity. To secure maximum returns from well bred animals the feeding and all features of environment must be made as favorable as possible to allow the exercise of the potentialities that have been intensified through generations of careful selection. The well bred animal is the only economical instrument for the person who wishes to realize upon his ability to feed skillfully and intelligently and care for domestic animals. Though feed is not the only factor of environment that must be considered in this connection, it is a principal one and many really meritorious animals prove a disappointment or fail to properly repay their owners because they were expected to perform the impossible and make bricks without straw. Perhaps this injustice is most commonly worked by under feeding, but the wrong kinds of feed are sometimes employed. Housing and exercise are important parts of environment, being allied to feeding in the same sense that they have much to do with nutrition and continued efficiency of digestion and the maintenance of health. On the whole probably more of our registered breeding stock is injured by too close housing than by exposure.

It is also imperative that this opportunity to develop be accorded the progeny of carefully selected parents during their growing days. It is very easy to allow a scarcity of the right kind of feed to continue too long, or to be deluded by the fact that an excess of fattening foods is filling the requirements because the animal is in an attractively fat condition. Often before it is realized, the days in which growth is possible have passed and a reliable knowledge of what was the animal's inheritance is impossible because no test was made of his capacities to respond to the demands which the builders of the ancestry sought to serve. In the future, much more than in the past, buyers

Feeding While Young.

of breeding animals of the meat-making breeds will want tangible evidence of the feeding qualities of their purchases such as can only be furnished by a record of feed eaten and gains made. Unusual prices of feeds required for growth sometimes justify a less rapid development than might be desirable, but generous feeding has the practical advantage of putting such indivduals as are to be discarded into their most attractive condition while still retaining the bloom of youth so that they can be disposed of to the best advantage. Referring again to the "corn crib cross," while it is an essential adjunct to heredity and enables the animal fully to utilize its inheritance in yielding maximum returns, it cannot originate any character or implant anything not represented in the inheritance. It must be relied upon as the material for the structure, the plan of which was drafted by the parents. It is only when an animal has been given a good chance to develop what it is supposed to have inherited that its value as a breeder, if a young animal, can be fairly estimated.

Of course there must always be considerable commerce in undeveloped and untested animals that are appraised upon the insight and experienced judgment of the buyers and sellers, but where much depends on an estimate of individual merit, as in selection of sires or additions to a breeding herd, a fair test must be regarded as superior to the best judgment.

To summarize the discussion of this factor in its relation to the individual, it may then be stated that economy of production suggests the furnishing of that environment most favorable to the development of those characteristics the animal is believed to have inherited. If difficulties may arise in so doing, then the best possible environment should govern the selection of the breed in order that heredity and environment may assist rather than combat each other.

It is as an aid to selection that environment is of greatest importance. Selection is based in large part on individuality and unless the environment permits the exercise of the inheritance, estimates of merit will be less trustworthy than is possible.

Good Care Aids Selection.

Then some inferior animals will be preserved and some superior ones discarded, and unsuitable environment will drag down instead of build up. It not infrequently occurs that offspring or descendants of animals of note are used as breeders without even having had a chance to come into a high state of development. In such cases the sole reliance is placed on the pedigree, and though it may be worthy of such entire faith at times, at other times it preserves what should have been rejected, thus misrepresenting the family and disappointing the owners.

The relation of environment to heredity in the development of individuality seems clear. The relation of environment to breeding powers is more difficult of understanding. Breeders sometimes state that this or that animal never had a chance to develop rightly, "but he is well bred and he will breed right." The likelihood of the transmission of the effects of environment is the question that has occasioned more debate and divisions among

Transmission of Effects of Environment.

men of science than has any other topic. Among biologists this question is designated, "The transmission of acquired characters." Just what constitutes an acquired character is hard to state, and much of the debate has been occasioned by a lack of agreement in the premises. The word "acquired" is used as opposed to "inherited," and anything acquired must therefore be the result of environment. It has been asserted that in the strict sense there can be no such thing as an acquired character because the most environment can possibly accomplish is the development of something of which the beginning is already present.* However, from the viewpoint of a practical stockman we will not miss the real value of the matter if we consider the transmission of development acquired as a result of environment.

Does the liberal feeding of the parents render their offspring any more responsive to good treatment than they would otherwise have been? Are the offspring of raced horses possessed of greater ability in the speed line than they would be if their parents were not raced? The supposed inheritance of the ability to perform certain tricks and to display certain habits are often introduced into this discussion but do not vary the principle of the cases here referred to. The transmission of congenital deformities and oddities are also offered as evidence in this connection, but must clearly be rejected because they have no relation to environment. In short, the matter resolves itself into the question, Does the hereditary material reflect the influence of the surroundings of the parents?

---
*Davenport, "Principles of Breeding," p. 358.

If we consider that all animal improvement has been effected under highly artificial environments, we can easily subscribe to the idea that the effect of environment is transmitted. It is when the attempt is made to give a physiological explanation of the occurrence that skepticism arises. According to the idea, which seems to be the best extant, that heredity is conveyed by tangible chromatin material retained in the reproductive organs, it is impossible to conceive of the actual incorporation within the chromatin of any substance representing the effects of special feeding, exercise, or education. Although biologists are still divided as to the transmission of acquired characters, they all regard chromatin as the chief if not sole vehicle of heredity. They therefore refer every matter to selection and seek to explain apparent transmissions of acquired characters or development on that basis. Not every instance can be satisfactorily explained by that means, but on the other hand no evidence is forthcoming to fully substantiate the other view. As a consequence a majority of the biologists, when pressed to give their verdict in the matter, have recourse to the Scottish jurors' "not proven."

Biologists on Transmitted Development.

Although comparatively few scientists approach problems of heredity to study their relation to stock-breeding practices, yet the breeders will find that the opinions of those less practical men are the most useful for explaining the best of what has been and is being accomplished in animal breeding. The denial of transmission of acquired development and the explanation of the effect of environment solely through selection may doubtless seem an extremely rigid adherence to the sufficiency of selection. This course is the only one open, however, to him who would put heredity on a tangible, truly scientific basis. To ignore or belittle selection is to ascribe results to the working of unknown forces and to continue an atmosphere of mystery around heredity that certainly can not achieve any advance in science or in practice.

It will be seen, however, that whether or not we believe in transmission of acquired development, there is no occasion to place a lower estimate upon good environment as related to improvement. The principal issue can be most satisfactorily discussed as it relates to trotting horses. With them there is no question about the acquiring of an unusual development, and whatever may be said will also apply in consideration of other effects of environment in other animals. The most plausible presentation of claimed facts presented of late years in support of the idea of the inheritance of acquired development is contained in the articles published by C. L. Redfield, based on his study of trotting horse breeding. He says:*

Acquired Development in Trotters.

"The theory relates to the inheritance by offspring of the characteristics acquired by parents. I have pointed out that the characters which an animal acquires are those which he develops by exercising them, and consequently that an acquired character does not mean the acquirement of a new character, but the development of a character already in existence. I have, therefore, substituted for

---
*"Horse World," Feb. 27, 1906.

'acquired character' the term 'acquired development.' I have also pointed out that a development acquired by exercise is in its nature dynamic, hence I have used the term 'dynamic development.'

"The next step was that if the dynamic development acquired by the parent is inherited by the offspring, then the amount of such should be proportional to the amount of the acquirement. This simply means that if the child is to inherit the dynamic development which the parent acquires, then the parent should acquire the development before he begets the child. Or to state the matter in another way, the child cannot inherit anything which the parent acquires after the child is born.

"I then pointed out that dynamic development is acquired by exercise, and as active animals continue to exercise during their whole lives, therefore, old and active animals have acquired more dynamic development than have young or inactive ones. In other words, I argued that the amount of dynamic development which an animal has acquired is a quantity to be determined by considering the age of the animal and the degree of its activity taken together. From this I drew the conclusion that if acquired dynamic development is transmitted from parent to offspring, then those animals which have, by natural inheritance, a fine dynamic quality must be descended from a line of progenitors which were either old or highly developed by special training.

"I have said that I took 1,000 registered stallions alphabetically, from the 'Index Digest' of the 'Register'* and calculated the ages of sires at the time when these registered stallions were foaled. From these I determined that the average time between generations in the male line was 10.43 years, which would give the average age of sires as 9.43 years at the time of service. I then said that, making all reasonable allowances for errors,

---

*"The American Trotting Register."

the average time between generations in the male line might be set down as between 10 and 11 years, and that this period might be used as a standard in testing the age part of the theory. So far no one claims to have tested the accuracy of my calculation; no one claims that the figures I gave were wrong; and no one has said that these figures cannot properly be used as a standard; yet if I am to be controverted, one of the first things to be done is to dispute the accuracy of my standard.

"I then took the entire list of 2:10 trotters as an appropriate class of animals to be used in testing the inheritance of dynamic development, and I calculated the ages of their male progenitors for four generations. The number of animals involved was over 5,000 and I gave the average time between generations in the male line for the production of 2:10 trotters as being approximately 14 years. This is an average of nearly 40 per cent over the standard average determined from the 'Register,' and my explanation of this remarkable difference was that it indicated the inheritance of acquired dynamic development. So far no one has disputed the accuracy of my computation and no one has attempted to give any other explanation of such an unusual divergence from the natural order of things. Am I right or am I wrong? If I am wrong will some one please come forward with a better explanation?"

Claimed Transmission of Acquired Development.

In the "American Naturalist"* the author has made this reply to the foregoing:

"It is noted that in the case of the average horses represented by the first thousand in the 'Index Digest,' the ages of their immediate sires only were computed, and found to average 9.43 years; whereas in the case of the

*Issue of January, 1909.

horses in the 2.10 list all the sires appearing in the first four generations were brought in. Assuming 14 years to be correct for the average time between generations, this carries us back 56 years.

"The first horse that was uniformly successful as a sire of speed was Hambletonian 10, foaled in 1849. In the sixties this horse's reputation as a sire of speed was established and he did heavy stud service until the time of his death in 1873. This was the real beginning of the trotting breed of horses. During the later years of the life of Hambletonian 10 and subsequent to his death his sons were patronized by owners of well-bred and speedy mares. The more successful of these were retained in service. When the grandsons of Hambletonian 10, with two generations of speed-producing sires back of them and out of selected female ancestry, came into service, it was found that in many instances they sired faster colts than did their sires or grandsire. Only in more recent years were representatives of popular families used for stud purposes in earlier life.

"In view of these facts, I deem it unfair to base a conclusion upon a comparison of two results, one of which (13 years as the average age at time of service of sires in four generations back of horses in the 2:10 list) comes largely from an investigation of the formative period of the breed, while the other (9.43 years as the average age at the time of service of immediate sires of average horses) mainly refers to more recent conditions. If the figures 9.43 and 13 had been derived by similar means their value would be unquestionable. A really fair comparison would demand the same procedure in one case as in the other. Either all sires in the four generations of the thousand horses should be used or else only the immediate sires of those in the 2:10 list.

"Assuming 9.43 to be correct for the average age of the sires when they produced the first thousand horses

DEVELOPMENT OF YOUNG STOCK

in the 'Index Digest,' I have attempted to secure a similar figure for the immediate sires of the horses in the 2:10 trotting list as published in the 'Yearbook,' Vol. 22. The list published in that volume contained 279 horses. In thirty cases the records failed to show the horse's age. In seven cases the age of the sire is not given. This leaves 242 of the 279 horses whose ages are shown.

"Below are given two extremes and the average for 242 horses regarding which there exists no uncertainty:

Horse.	Foaled.	Sire.	Sire, foaled.	Age of Sire at time of service.
Wentworth, 2:04½	1903	Superior	1879	23
Dolly Dillon, 2:06½	1895	Sidney Dillon	1892	2
Average for 242 horses............				9.41
Average age of sires of 2:10 horses given by Redfield......				13
Average age of sires of average horses............				9.43

Of the 242 horses, 1 was sired by a 2-year-old stallion.
      11 were " " 3 " " "
      17 " " " 4 " " "
      30 " " " 5 " " "
      19 " " " 6 " " "
      21 " " " 7 " " "
      21 " " " 8 " " "
      25 " " " 9 " " "
      14 " " " 10 " " "
      17 " " " 11 " " "
      8 " " " 12 " " "
      13 " " " 13 " " "
      8 " " " 14 " " "
      9 " " " 15 " " "
      6 " " " 16 " " "
      6 " " " 17 " " "
      1 was " " 18 " " "
      4 were " " 19 " " "
      3 " " " 20 " " "
      0 " " " 21 " " "
      6 " " " 22 " " "
      2 " " " 23 " " "

Inheritance Not Related to Sire's Age.

"Taking 9.43 years as the average age of the sires of average horses and substituting 13 by 9.41 years as the average age of the sires of 2:10 trotting horses, it is evident that the records do not reveal any superiority of the old sire over the younger one."

It seems quite possible and fully reasonable to account for the great accomplishment of breeders of American trotting horses without accepting the idea of transmission of acquired development.

Most fast horses come from parents with speed because our most astute breeders insist on actual performance as a test of individual merit. Sires without low marks are seldom accorded the opportunity of choice mares until their get from mares bred to them earlier have demonstrated their possession of speed. In the natural order of things such a sire must be of considerable age before being used very freely and having many colts, and thus it might seem that age in the sire favors speed in his get. Whichever side of the question a breeder chooses to take his practice will not be seriously changed. In one case he will train his breeding stock as an aid to selection and discard the failures. In the other case the training will be calculated to produce a result transmitted to the progeny. Those not responding to the training will be discarded just the same.

How Development Aids Selection.

The development of such a quality as early maturity or ease of fattening may likewise be regarded as not attributable to transmission of effects of environment. Such a character being desired, and environment adjusted to develop it, those not showing it are eliminated while those with greatest aptitude in the desired direction are mated and the inheritance made greater in some of the offspring than it was in either parent. The same sort of testing and selection continued for several generations tends to render the inheritance pure to the desired character. It would be out of line with the practices of master breeders as well as scientifically wrong to seek to develop and have transmitted any greater degree of merit in any feature than was inherited, but selection can, by mating two suitable animals, procure for their offspring a more generous inheritance than was possessed by any individual of earlier existence.

The light esteem placed upon the possibility of transmission of the effects of environment should not incline us to in any way lower our appreciation of studied care and intelligent liberal feeding of breeding stock. Whether we think of speed, easy fattening or natural fleshing qualities, we see that it was only under favorable environments that the makers of the breeds and strains were enabled to select from their herds those animals that should be mated for the perpetuation and intensification of the features they sought to impress upon their stock. It is only by continuing the same conditions that we can retain or improve those same features. Those conditions being withdrawn we are forced to rely entirely upon pedigree, which though of the best cannot safely be allowed to overbalance individuality, and good individuality necessitates full chance for development.

Actual Role of Environment.

It is only when improved animals are subjected to scrub conditions that the pull of environment is downward. The maintenance of a favorable environment in the feed particularly is therefore necessary in order to realize upon the good inheritance of the animals. It is again essential in the selection of the really best individuals from the young produced. If environment is properly related to the purpose sought its pull is upward. To maintain our flocks and herds under an increasingly artificial environment and yet retain in them the vigor and freedom from difficulties of reproduction found in native stock is the duty that falls upon the breeders of today and their successors.

CHAPTER XIII.

DETERMINATION OF SEX.

Probably no other one thing has occasioned so much speculation regarding the wonderful processes of reproduction as has the desire to control sex. The power to do so would be temporarily at least very profitable to breeders. Their general desire to have the increase of their herds consist mainly of males or females, according to which would be most profitable at the time, is responsible for a multiplicity of directions for ensuring the production of male or female at will.

In the human race the sex of the foetus is distinguishable at about the eighth week of pregnancy. It is not known whether the sex is determined when the spermatozoon enters the ovum or whether ensuing conditions are responsible for the development of male or female organs. The fact that sex is discernible only at the eighth week by no means indicates that it was previously undetermined.

The commonest idea about sex determination is that females bred at the beginning of the period of heat produce male offspring. Other notions are based on the same supposed principle, namely, that an ovum fertilized while immature produces a male; maturity is supposed to be in propor-

Influence of Time of Breeding.

tion to the age of the ovum and the nutritive condition of the dam.

Fertilization ordinarily occurs in the Fallopian tubes, the ovum descending from the ovaries when fully prepared and separated from its containing sac through which alone nutrition can be supplied. Early or late service can therefore have no connection with the completeness of the nourishment of the ovum. As to the possibility of sex being attributable to a fresh or stale condition of the ovum, it is inconceivable that anything so fundamental as the production of male or female organs could be controlled by any change resulting from a few hours' residence outside the ovaries but still subject to internal body conditions. Furthermore, if females were the result of service late in the period of heat, then in herds and flocks continuously accompanied by males, and where service ordinarily occurs at the first indication of heat, we should find all male offspring. The experience and observations of managers of large cattle and sheep ranches do not substantiate this idea. Evidence of heat and the time of service are at best crude indications of the real time of fertilization.

Considerable publicity has been given the theory of Dr. Schenck, whose advice it is purported has been sought by royal families of Europe. Schenck supposes sex to be influenced by the condition of the ovum at fertilization. When the urine shows a large proportion of sugar he argues a lower nutritive state of the body and therefore the unripeness of the ova because of the less amount retained, and con-

Influence of Body Conditions.

ception occurring at such time must result in a male. No extensive statistics covering tests of Schenck's theory are available; inasmuch as in any case there are equal or slightly greater probabilities of the production of male offspring, no surprise need be occasioned by isolated instances of the appearance of males succeeding endeavors for their production. The long-continued practice of flushing ewes at mating time has never been claimed to influence the sex of the lambs as would seem to be the case if this idea were correct, though it is fair to state that Schenck emphasizes the composition rather than the amount of food.

Alternating Ova. It was once held that the right ovary produces germ cells that always result in females, while those from the left are male, but instances are now known where females with one ovary removed still continue to produce offspring of both sexes. A similar claim was made with regard to male parents, but experiments conducted by James Buckingham of Zanesville, O., disproved this. Mr. Buckingham used nine sows divided into three similar lots. In each lot the first sow had the right ovary removed, the second the left, and the third was normal. One lot was bred to a boar whose right testicle was removed, the other two to boars with the left one removed. The litters had from seven to nine pigs each. In no litter was there less than three males or more than five females. This experiment, reported by Mr. Buckingham in the "Country Gentleman" of 1865, shows clearly that neither ovary or testicle produces either sex exclusively. An idea discussed in the foregoing is the basis for the notion which sometimes finds expression in the direction to breed an animal in the first, third or fifth period of heat after the delivery of a female if a male is desired. The ovaries may act alternately but there is no reasonable ground for supposing that they differ in the sexual possibilities of their ova.

Some believe that the older or more vigorous parent will control the sex. This would seem to suppose that the hereditary material represents the sex of the parent from which derived and that in development the supposed greater activity of the stronger parent's germ-plasm dominates that of the other and directs the production of the organs of the corresponding sex. Why this should be so only in regard to the formation of the sexual organs is hard to see. Neither have we any good grounds for supposing that the germ plasm represents or seeks to produce the sex of the animal from which it was derived. That such is not so in bees has been shown. The unreasonableness of this theory and the lack of data to support it render it untenable.

Influence of Stronger Parent.

The nutrition of the pregnant dam has also been thought to have a bearing on the sex of the offspring. As was previously stated, the fact that in higher animals the sex of the foetus can be observed only toward the close of the first quarter of the period of gestation by no means indicates that conditions prevailing subsequent to fertilization control the sex. On the other hand it is not proved that such is

Effect of Nutrition.

not the case. The general assumption is that the development of female young demands greater amounts of food and more favorable conditions than does the production of males. The claim is made that statistics reveal the fact that in countries that have been ravaged by war, and the food supply of the inhabitants diminished, an increased proportion of male children is noticeable. This is presented as a natural provision for the restoration of the proportion of males depleted by the conflicts. Data compiled by other students of the subject do not show any disproportion of sexes attributable to nutrition. Considerable experimental evidence is available on this point. Although relating to lower forms of life this evidence may fairly be considered as bearing upon the principles that govern sex in domestic animals.

Experimental Evidence.
Born (1881) reared 1,443 tadpoles on a highly nitrogenous diet and secured 95 per cent females; others on ordinary diet gave 62 per cent females. While the tadpole is an independent animal its transition to the frog stage corresponds closely to foetal development of larger animals. Born's method of observing the sex is now claimed to have been inexact. Yung (1883) reported tests in each of which a liberal nitrogenous diet was furnished developing tadpoles, giving in each case over 70 per cent females. More recent repetitions of the same test have failed to make it appear that the food influences the sex of frogs. Cuenot found great irregularity in the proportion of females in separate tests, sometimes males predominating and sometimes females. Much care was exercised in de-

termining the sex, although as in earlier cases the sex of those dying during metamorphosis could not be ascertained. On the whole he considered that his results did not indicate any influence of nutrition upon sex. Cuenot also experimented with rats, one lot being liberally fed with a variety of food materials while the other received only limited quantities of bread. The first group produced forty-three males and forty-nine females. In small litters it would seem that each individual would be fully nourished and according to the nutrition theory an excess of females should appear. The total for all litters comprising less than nine young were, seventy-one males and sixty-two females.*

From 1867 to 1873 three investigators reported results favoring the idea of supposing that nutrition has an influence upon sex in butterflies. In one case Mrs. Treat reported that a starved lot produced thirty-four males and one female while a well-fed lot gave four males and sixty-eight females. Five other experimenters reporting between 1868 and 1874 found no such influence. It was pointed out by Riley that since females are weaker a greater number of that sex would succumb under adverse food conditions, thus showing a preponderance of males among the living ones produced, not through the influence of nutrition upon sex but because of the elimination of the females before emerging as adults to be counted.

Experiments with Butterflies.

Kellog and Bell of California have made an exhaust-

---

*Discussion of sex in Morgan's "Experimental Zoology," chapter 25.

ive study of this subject. Their results have to do not only with the effect of nutrition upon the developing silkworms with which they worked but also show the influence upon sex of liberal and scant, or as they term it, optimum and minimum food supply, furnished to parents and grandparents. In their table which follows M indicates minimum and O optimum food supply. The number of deaths before maturity are separately listed. An examination of the table will show that the five lots receiving a minimum diet produced fifty males and forty-six females while the three lots on optimum diet produced twenty-three males and twenty-seven females.

EFFECTS OF OPTIMUM AND MINIMUM FOOD SUPPLY.

Lots.	Fed.	Parents.	Grand-parents.	Deaths before maturity.	Males.	Females.
1	O	O	O	2	13	10
2	M	O	O	2	14	9
3	O	M	O	3	8	14
4	M	M	O	6	8	11
5	M	O	M	0	15	10
6	M	O	M	0	11	14
7	O	M	M	20	2	3
8	M	M	M	21	2	2

Argument in favor of nutrition as a determinant of sex is sometimes based on the development of the queen bee. It is known that when a colony loses its queen the workers by furnishing liberal amounts of the royal jelly to a larva develop for themselves a new queen. This larva would otherwise

The Evidence from Bees.

have become an ordinary worker. If we could properly assume the worker to be of the male or neuter sex it might seem that the extra food was responsible for a change in sex; the worker, however, is in reality an undeveloped female, and the effect of the extra food utilized by the queen was to develop in her the egg-laying organs and make her a functional female. The nutrition effects no change in sex for femaleness was already produced. The nutrition permitted the completion of the egg-laying organs. Apparent evidences of the effect of nutrition upon sex, as is seen, lend themselves to other interpretations and cannot be entertained by a fair-minded practical breeder. In the interesting case of the queen already referred to it is well known that the eggs she lays prior to her maiden flight and impregnation produce males exclusively. Subsequent to impregnation all offspring are workers, or undeveloped females, any one of which presumably might become a queen if properly fed during development following hatching from the egg. In this case then it seems apparent that sex was determined in the fertilized egg. The fact that the unfertilized eggs gave males and those fertilized gave females would go to indicate that here the parent transmits its opposite sexual qualities. If it be really true that the matter of sex is settled at fertilization and that this applies in domestic animals, it must dissipate all hopes of our ever being able to control sex.

Sex Probably Determined at Conception.

The production of twins constitutes the only piece of evidence discussed as bearing directly on the point in larger animals. In the human race and in cattle the bearing of twins is the exception to the rule and twins are sometimes no more alike than children of the same parents born at separate times. In other cases the resemblance is so great as to render distinction very difficult. Like or identical twins are believed to be the result of the separation of the two cells produced by the first division of the fertilized ovum, the two halves developing separately producing two individuals with exactly similar inheritance. Unlike twins are considered to be produced by the fertilization and development of two distinct ova. As will be readily recognized in this case the inheritances may differ very widely. Since the like twins, those produced by a single fertilized cell, always have the same sex, it seems fully probable that the germ cells themselves contain the determinant of sex and that it is not dependent upon conditions governing gestation. We have also dismissed theories of control of sex based on ideas presented in the first section of this chapter. We have more scientific and more reasonable grounds for considering sex to be determined when the reproductive cells unite.

Further evidence to support the thought that sex is determined at fertilization is drawn from recent investigations of the germ cells of a number of species of insects. Recently several investigators have located extra chromosome-like bodies in germ cells of numerous kinds of insects. These extra bodies, or accessory chromosomes as they are now improperly called, were not at first regarded as chromo-

The Accessory Chromosome.

somes because of their unusual size, being sometimes larger and sometimes smaller than the ordinary chromosomes among which they occur. Further difficulty was also afforded by their having no mates.

The accessory chromosome was first noted by von Siebold in 1836; he found it only in the spermatozoa of a particular snail. Following this there were found several instances in which spermatozoa were of two equally numerous kinds, those having a certain number of chromosomes and those with one more than that number. In the early stages of male germ cells this accessory chromosome splits and divides in the ordinary manner. In one of the reducing division stages it fails to split, however, thus producing from each spermatocyte two ordinary spermatozoa and two with an accessory chromosome. It was suggested by McClung that when the female egg is fertilized by a spermatozoon containing the accessory chromosome the resulting offspring would be of one sex while the union of the ovum with an ordinary spermatozoon would produce the opposite sex. Shortly afterward however a similar irregularity was found to exist in the eggs of some females.

Significance of the Accessory Chromosome. It is clearly established that in the case of the common squash bug (anasa tristis) the body cells of the female have twenty-two chromosomes and those of the male twenty-one.* The same is true of the primitive germ cells. In the reduction of the germ cells with twenty-two

---
*Wilson: Studies on Chromosomes, "Journal of Experimental Zoology," II, 1905; III, 1906.

chromosomes all ova will of course have eleven chromosomes each. In the case of the male germ cells, however, since the twenty-first chromosome does not divide, one-half the spermatozoa will have ten and the other half eleven chromosomes. If, then, a eleven chromosome egg be fertilized by a ten chromosome spermatozoon the offspring will have twenty-one, the number occurring in males. If it be the eleven chromosome spermatozoon that fertilizes the ovum, the resulting number will be twenty-two or female. The same explanation is repeated in the diagram:

```
Spermatazoon.                                  Egg.
  10————————————}———————————— 11 — 21 — male.
  or                                            or
  11————————————}———————————— 11 — 22 — female.
```

The known facts concerning the accessory chromosomes by no means dispose of the problem of the determination of sex. The fact, however, that the male and female adults in the case referred to have different numbers of chromosomes in their body cells is practical proof that the accessory chromosome is associated with sex, and this being true, the happenings presented in the diagram are highly probable. This is additional strong evidence that sex is determined at fertilization, and that it is beyond human influence before so determined. In

Undesirability of Sex Control. view of the present study of the subject it seems quite likely that we may soon have a more intelligent idea of the basis of sex. That sex of farm animals should generally be under the control of man seems hardly desirable. The present nearness to equality of numbers of the two sexes is of great importance in the preservation and improvement of types by affording a large number of males from which to select the best ones for service. Full control over sex would seem to give man a power that would not be exercised to his own best advantage or to that of the races of his animals.

CHAPTER XIV.

FOUNDATION AND MANAGEMENT OF A BREEDING BUSINESS.

The preceding pages have not dealt with the origination and perpetuation of new types or characters; that phase of heredity is reserved for a later place. Discussion has been confined to those aspects of heredity most likely to suggest why things are as they are, and how existing types and characters of excellence may be rendered most nearly certain of reproduction. In a sense it may be truly said that breeding is entirely based upon the single principle of selection; that is, if selections are right the desired results must follow. Competent selection, however, has been shown to be dependent upon diverse considerations and may be said to necessitate the application of the further principles of judging, feeding and other requisites of development, and each of those in turn amounts to an independent study.

It is obvious that stock-raising cannot become an exact science. It consists of the handling of the hereditary material that does not lend itself to direct examination or to manipulation. The breeder's calling must be regarded as an art. Extensive studied experience may in part be substituted

Breeding an Art.

by study of allied sciences, yet the chief factor is the natural personal equipment of the artist or breeder. The use of tests and records at first blush may appear to give the work a mathematical aspect, yet as was shown, the only positive feature it introduces is the measure of individual functional capacity and leaves much else to be considered in mating.

An understanding of the relation of selection to breeding and of environment to selection does not suggest any procedures different from those of the best breeders. It does afford an appreciation of the necessity of unflagging attention to each detail in each animal's life. Although this is an asset that can only be fully acquired by actual experience a large measure of it may come as a result of a study that gives familiarity with the facts and a working appreciation of the practices of those who have succeeded. The personal qualification of first importance is an intelligent liking for the work. It is easy to allow the glare of the showring to convince one that he would enjoy the stockman's life, but a better test of inherent fondness is a close contact with the details of daily care, not so much of charges destined for showing as with the breeding stock and the young animals in the making. Such associations will not originate fondness for the work in persons talented altogether in other lines, but where there is natural inherent liking for animals the proper environment will develop useful and reliable leanings and opinions. With this fundamental equipment assured the would-be breeder is likely to apply him-

The Breeder's Personal Equipment.

self in such a way as to reach a fair degree of efficiency in the special branches of his calling.

The first special branch is that of judging animals and judging pedigrees. Stock judging is so widely taught that practically anyone can receive the aid of the teaching of persons experienced, not only in judging but in the teaching of judging. Practice alone, however, can give real efficiency in judging. It is essential to recognize that the basis of judgment is the faculty of making a ready and fair decision upon each animal as an individual. One who has not rightfully acquired sufficient confidence in his own judgment to be ready to retain his opinions in the face of some opposition is still unprepared to render decisions of moment to himself or others. The attractiveness of the work lies in the fact that with most animals there can be no practical test of the correctness of any judgment. The only criterion is the opinion of others, and the safety of the calling lies in the fact that there is no one single standard of perfection and not all breeders can be pursuing a wrong ideal at one time. Ability to officiate in the show-ring is not a guarantee of ability to do all the judging a breeder must do. An important part of a breeder's judging consists in passing upon the merits of animals he already owns or has raised and to properly estimate their merits in comparison with those owned and raised by others. The disposition to allow personal ownership to blind one to the defects

Judging Ability.

Impartiality Essential.

of his stock, or to fail to recognize the merit in that of other breeders, is an insurmountable barrier to improvement.

The judging of pedigrees demands acquaintance with at least the more recent history of the breed handled. This requires application and a retentive memory. One who has the natural interest in the work will find his memory capable of retaining a fair knowledge of pedigrees for at least one breed, even though lacking in other matters. A full knowledge of breed history, including the practices of the founders and improvers, their conditions and the demands made upon them, gives a breeder an intelligent understanding of and wholesome respect for his animals. It also gives a greater fitness to judge the future of his breed and its relation and his own relation to agricultural progress.

Value of Breed History.

Feeding is a chief factor in environment. It constitutes a separate study which must be mastered and applied. An active interest in each phase of the work is also necessary in care and feeding to insure the proper observation of the peculiar wants of each individual and the adjustment of the environment to those peculiarities. Only the most judicious care and feeding can render possible any more than a rare chance realization of the possibilities the animals have inherited.

Breeding is a many-sided profession. A man may have and do all that has been suggested and yet fail of accomplishment through poor salesmanship. Ordinarily speaking, a breeding enterprise once launched

Salesmanship.

SHOW HERD OF SHORT-HORNS.

must be its own financial support. New males and additional females must be of distinctly high class and will cost proportionately. Their purchase price must be raised by the sale of a part of the increase. No matter how well bred and how well developed the increase may be, some skill is required to dispose of it advantageously. The principles of selling live stock are exactly the same as those of selling any other commodity. The production of a high-class article is the first step. Then it must be brought to the attention of the people for whom it is best suited. Advertising through the press, the show and through sales must all be studied. The buyer having been reached, it must be recognized that the treatment accorded him is sure to be either a good or bad advertisement. The disposition to fairly recognize the interests of the buyer on every point is the basis of satisfactory dealing. The raiser of market stock has the great advantage over the breeder of pure breeds in being able to sell wholesale at actual values at any time. In either case, however, the buyers usually seek the seller when the highest class of stuff is wanted and of necessity permit him to make the prices; with inferior stuff the reverse is true.

Advertising.

The chief personal qualification of a breeder has not yet been touched upon. It may be spoken of as the courage of convictions. No matter how complete may be the knowledge of what should be done, unless there is a practical faith of sufficient strength to carry the information into execution, no re-

Executive Ability.

sult can be obtained. Of breeding it is as true as of other professions that the number who have fallen because they failed to do what they knew should be done is much greater than of those who fell short of the top through not knowing what to do.

Opportunities for profit and distinction sometimes attract men with large financial resources to the profession of breeding. But these may lack the true stockman's personal inheritance. The early failure of such men is cited by the uninitiated as illustrating the caprices of the calling. On the other hand it is a fact that much of the most valuable service to the stock-breeding world has been and is being performed by men whose large means were accumulated in mercantile lines. Undoubtedly the financial standing of such breeders is of prime assistance in enabling them to bring together animals whose mating produces merit not before attained; matings which would have been impossible to persons of more modest fortune. An equally vital factor is found in their business training. Such men have already demonstrated their business capacities in accumulating the means they choose to employ in breeding. The business principles of doing everything at the right time and undertaking nothing that cannot be thoroughly done are as productive of returns in stock-breeding as in any other field. It is not necessary to minimize the artistic side of breeding in order to emphasize the dependence of the proceeds and continuation of the work upon the observance of the best principles that govern production and sale in all lines. Use of poor

Wealthy Breeders.

materials and indifference or lack of study in their combination may yield an occasional chance article of value, but such a system can lead only to disappointment.

Conservative procedure in establishing a breeding enterprise will avoid many serious handicaps to its continuation. Location has much to do *Location.* with breeding, with rearing and with selling. So far as the selling is concerned the prospective trade in the immediate vicinity is really of secondary importance. Though it may be less true in the future than it is now it is a fact that a large proportion of buyers of breeding stock are more appreciative of animals reared in remote parts. Facilities for selling afforded by a given location may be chiefly considered with regard to ease of access for buyers, convenience of shipping and proximity of other breeders.

More important still is the matter of securing a location that will allow at lowest cost the environment to which the breed has been accustomed and for which it is calculated. Even though no dependence is to be laid upon the demands of the immediate vicinity, yet it would be overweighting the venture to attempt to produce stock for other sections under conditions not naturally favorable. Over-enthusiasm with the thought of what it is hoped to produce sometimes hinders the wisest choice of location. If selection be right and the surroundings fail to furnish the requisites for development, so far as feed is concerned, reliance can still be had on purchased materials. Purchased dry fodder, grains and by-products may be as well for the animal as similar home-raised feeds but cannot supplement rich pastures and fresh green fodders in the fullest and most economical development of any class of stock. A small percentage of outstanding animals sell at a price out of all comparison with their cost, but the majority, even including some of the really useful ones, find their market at a figure not very far from their cost under reasonable conditions. In other words, the margin of profit in the majority of instances is not a very wide one. If a record be kept of all items of expense in producing any animal it will be found that the outlay for feed exceeds any other item. The difference between cost of purchased and home-raised feeds may easily amount to as much as the difference between a profitable and a losing selling price on the animal, even if it were possible to secure the same development as might be had with crops always ready at hand and in most desirable condition for feeding.

Home Grown Feeds.

It is rather exceptional, however, that a breeding enterprise is planned before the consideration of a location. In a majority of instances a breeding enterprise evolves from a desire to secure a product of greatest value from pasture or feeds available under a fixed set of conditions. The conditions being already set, they must govern the selection of the class of animals to be dealt with. In America to-day one sees leading breeds of draft horses and of beef cattle competing with each other for popular favor in counties where the conditions are uniform, the systems of rearing similar, and all raisers catering to the same

Strain More Important Than Breed.

market; there is apparently no recognition that each breed is a product of selection for the special requirements of peculiar sections. When one also witnesses adherents of the same breed differing with each other as to which particular type should be preferred he may well be confused. It is necessary to understand that while pronounced features of utility are uniformly characteristic of breeds, yet peculiar and inexplicable personal likings have caused persons aiming to fill the same want to select different breeds, and the raiser of one breed may be selecting and feeding for a standard very different from the one that inspires another adherent of the same breed. It is not sufficient to know the characteristics, history and accustomed environment of the breed in its home surroundings; the same study must apply to the separate strains within the breed.

The most practical method of becoming a seller of breeding stock is to commence by breeding for the market. Practically all the founders of existing breeds were at first in this position, but since at present registered animals must be descended from other registered animals, some such stock must be included in any herd that is not expected to continue to supply regular market trade. Such added pure-breds should of course be of sufficiently high order of merit to give promise of improving the rest of the herd. The price at which they may be bought however must be in proportion to the value of the excellence they can impart to their progeny.

Starting from Market Stock.

In draft horses and meat-producing breeds of stock

the real feeding and marketing of all male increase show conclusively which are the most profitable dams and they can be retained as proved individuals together with their female increase, which in their turn may be likewise required to prove their right to be retained. Such selection, based only upon demonstrated merit, renders possible the accumulation in course of time of a band of females of highest value. It necessitates of course the use of only such sires as promise to improve the best that has already been attained, and it implies the immediate elimination of any whose get are not superior to the dams. Such management, while not the fastest route of entry into the breeding fraternity, may be very fruitful both of financial reward and of experience, and it certainly gives the best possible foundation.

When it is planned to enter immediately upon the scheme of selling the young stock as breeders, the selection of the foundation stock is a test of actual faith in the principles and standards. Animals that are at least fully equal to what might be procured under the plan just discussed should be chosen and will of course come at high prices. The best is none too good to start with and should be insisted upon even if funds demand restriction of purchases to a single individual. The fact that an animal is priced high is not a guarantee of merit, nor is a low price proof of inferiority. Unless the beginner is sufficiently experienced to be able to procure a fair return for his expenditures purchasing had better be postponed. Young breeders of limited means are sometimes counseled

Not How Many, But How Good.

to select a number of animals of ordinary merit in preference to a smaller number of superior ones of equal value. The assumption is that experience can be acquired with less risk of loss and improvement can be effected subsequently. If it planned to place such stock in commercial herds and offer nothing for sale for breeding purposes, the counsel is well given. If, however, the immediate increase is to be retained or offered for sale as breeders, acceptance of such advice is a step toward disappointment and failure. The difficulty of disposing of the offspring of mediocre parents and the meagreness of the profits they return are very unlikely to encourage the subsequent addition of really good individuals. The common character of such stock and the lack of interest they inspire are almost sure to result in lack of the care and opportunity they especially need and which would be more readily accorded to more attractive and more highly prized individuals. In the desire to multiply numbers worthy and unworthy produce are likely to be retained and the enlargement of numbers is the chief if not the only direction in which progress is effected.

Cheap Foundation Stock.

"The best is the cheapest" is a thoroughly practical maxim. The offspring of the female of common individuality or common ancestry, or both, requires as much expenditure of care and feed to bring it to a salable age and condition as does an outstanding good one, and it is in turn sought only by those with whom a low price is a primary consideration.

Merit in Both Parents Essential.

HEREFORD SHOW CATTLE.

It is true that lack of pronounced merit in foundation females can be largely counterbalanced by extraordinary excellence of the male with which they are mated, but real conservatism would favor a small equipment with the basis of success laid on all sides, rather than a more extensive but unbalanced plan of which one part is expected to atone for the other, instead of placing reliance on each. "Not how many but how good" is no more applicable at any point than in the laying of the foundation. Though it will of course not be feasible to purchase any number or even a few females of the highest approach to perfection, uniformity of type among what are procured must first of all be insisted upon. Such defects in conformation as really cannot be avoided must find their counteracting force in the male. Obviously, then, it is desirable to postpone the purchase of a male until the females are on hand for comparison and study.

It is a severe test of any sire to be called upon to stamp his excellence on the offspring of females which even though they be uniform in type are gathered from different sources and represent different lines of breeding. Only an impressive animal, strong where the females are weak and with unusually good ancestry, can be relied upon to meet such requirements, and as is commonly said he should first be selected and then purchased.

The First Sire.

The use of any unproved sire is somewhat of an experiment and the greatest danger lies in failing to recognize and admit that such a one is not leaving offspring as good as they might reasonably be expected to be. The best values are sometimes offered in successful sires owned by men who insist on changing to avoid in-breeding, or to avoid keeping two males. If a well preserved though aged male that has proved good is obtainable, no objections can be raised to justify passing over such a one for the most promising young and untested individual. It is in the selection of a second sire that greatest difficulties are likely to arise. Even though the first should be satisfactory, his progeny are soon ready to be mated and unless the number of females on hand justifies the maintenance of two sires a new one is necessary. Undue haste is often exercised in disposing of the old sire. If he has been successful in improving upon the merit of the females he may well be retained until the experimental stage of his successor's career is passed. The trouble and expense of maintenance is but small in comparison to what is gained, and there is the further advantage of adding to the number of females of the same breeding. This similarity of breeding of the females renders the selection of a later sire much more simple than it is when he must be mated with females of varied inheritance. The selection of the second sire is more difficult than that of the first because he must be suited both to the remaining original females and to their offspring. If these latter are an improvement upon their dams, as they should have been if retained, then the second sire must be of considerably higher character than his predecessor. It is not a wise procedure to make the experiment of mating all the females to even the most promising young male. He should be procured some time before he will be most needed and mated with older females whose breeding qualities are known. This affords a means of detecting objectionable features in his get without allowing him to mar the entire increase of a breeding season.

Phenomenal Sires. Many a breeder has brought his herd into prominence through the merits of a single phenomenal sire. The possession of such a phenomenon is in considerable degree attributable to good fortune, for not even the most discerning can positively say that this or that unproved male will beget progeny uniformly of extraordinary merit. Nevertheless he who is most assiduously in search of and determined to procure a herd-header of unusual rank, and who has the faith that will prompt him to secure possession of the nearest approach to his standard, is the one to whom the good fortune of the phenomenal sire is most likely to fall.

Even if no serious mistakes are made in selecting sires or in culling of female increase, it takes a good many generations to reach a satisfactory degree of uniformity in the excellence of an entire herd. *Culling the Females.* Not all can have an inheritance of unmixed good, and the weaker ones must be weeded out as they appear. A sire may leave some undesirable character in some of his get which it may be deemed wise to retain for other reasons, or some of the more valuable families may be less prolific than is necessary to enable their increase to dominate the whole herd. With continued progress new and higher standards come to be held. Perfection or even a close approach to it in every individual animal is not to be expected, and though it were attained, there must still be variation in the abilities of those animals to reproduce their excellence. A breeder may be said to have achieved a measurable degree of control over heredity and to have earned distinction when the most of the animals he rears can be relied upon to elevate the herds of his contemporaries.

Such a height is not likely to be reached without considerable showyard fame attaching to the herd, though the estimation accorded the stock by those who have found it a dependable source of desirable qualities is of even more moment than ribbons won in competitions where individuality alone must govern. Such uniform excellence and prepotency can not be expected within the second or the third generation from even the choicest band of females brought together from different localities and which must necessarily be dissimilar in breeding. It is when the female ancestors have been for many generations under the immediate direction of a man who is capable and actively interested that the possibilities of selection appear to the best advantage, and the greater the length of time such condition has obtained the greater the possibilities. The real foundation herd must be bred. Such a history allows the owner or manager to have within his own memory a personal acquaintance with each ancestor of likely influence. It enables him to know fully all inherent tendencies and to dictate the matings with the greatest possible measure of assurance. Mistakes are sure to be

Uniformity in Females.

made and reverses experienced and these must be corrected and overcome. Though much may be accomplished in one or two decades, even the most capable breeders will find the fruits of their selection becoming more and more apparent in proportion to the number of years of unremitting vigilance and application. In this lies the most important real advantage of European breeders. Their herds in many instances have been handed from fathers to sons, who have continued selection toward the same ideal until the hereditary material has been freed from all serious impurities and represents a degree of strength and purity that insures the highest degree of excellence and prepotency.

Value of Long Established Herds.

The animals of an old herd have pedigrees in which the owner appears almost to the exclusion of other breeders; in fact he has constructed the pedigrees of his animals and in proportion as he has succeeded and has earned the confidence and respect of his fraternity the succeeding breeders will prize the results of his labors.

CHAPTER XV.

INBREEDING AND LINE BREEDING.

In the minds of the majority of persons who consider matters relating to breeding, the idea is firmly fixed that the practice of inbreeding is altogether objectionable. In general it is described as the mating of close relationships, though what is sometimes considered to fall within the term is in other instances designated by milder phrases. In really correct usage inbreeding designates the union between brother and sister or between offspring and parent, in one or more generations. Popularly considered line breeding is applied to matings of a degree of consanguinity not included in the foregoing definition. Those who advocate breeding in line fear the results of actual inbreeding, and aim to avoid the uncertainties of mixing strains or families totally unrelated though belonging to the same breed. In the minds and speech of present-day breeders, however, there is no generally recognized distinction between inbreeding and line breeding. The principle involved is the same, and the difference is one of degree. Most of what is commonly referred to as inbreeding is not in reality such. In order to be clear in discussing the subject as it is before our breeders today inbreeding will be considered as the mating of two animals with suf-

Inbreeding Defined.

ficient similarity of ancestry to make 50 per cent of the inheritance of one identical with the same proportion in its mate, the inheritance in common not being necessarily that of a single animal.

TWO PEDIGREES SHOWING INBREEDING.

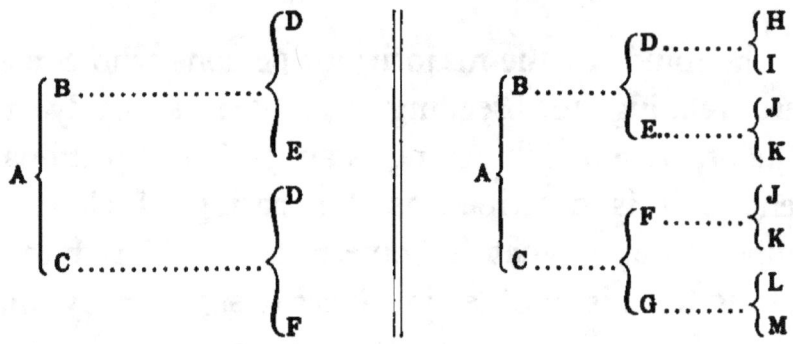

In each of the above cases A is inbred. In the first instance son and daughter of D are mated. In the second, the dam of B is full brother to the sire of C, that is, B and C are full first cousins.

Line breeding may be differentiated from inbreeding by defining it as the mating of two individuals identical to the extent of 25 per cent and less than 50 per cent of their blood. Line breeding however really implies something more; it implies a succession of sires that trace their descent in some measure to the same individual. The example on the opposite page shows typical line breeding.

Line Breeding.

In this case the descendants of the female I have been line-bred. G having two lines to I has been mated with F who is also a grandson of I and their progeny C is bred to B, another grandson of I. In no case have full first cousins been mated. The blood of I preponderates in A and yet there has been sufficient latitude to allow the selection and use of only the better individuals from the descendants of I, and so valuable characters may be thus retained, not so fully as might be if inbreeding were practiced but in a safe and useful degree.

AN EXAMPLE OF LINE BREEDING.

```
        ┌ B ┌ D ········ { I
        │   └ E ········ {
        │       ┌ H ········ { I
A ┤     ┌ F ┤
        │   └ J ········ {
        └ C ┤
            │   ┌ K ········ {········ { I
            └ G ┤           └········ {
                │   ┌········ {
                └ L ┤
                    └········ { I
```

The difference between line breeding and inbreeding is one of degree. The principle being the same it may be discussed at once for both. The popular use of the term "blood" in this connection is likely to be misleading and should be substituted by some term to designate the hereditary material itself. Also there is strong probability of error in assuming that because of the parentage mentioned that 25 per cent of the germ-plasm in B is identical with a similar proportion of the same material that entered into the make-up of C. The process typified in maturation shows that there may be a wide difference between the combination of chromosomes received, for instance, by E from J and K and that received by F from the same individuals in the sample pedigree showing inbreeding. This possibility of dissimilarity may explain the variable results observed in apparently similar instances of the mating of closely related animals.

Most states have laws forbidding the marriage of first cousins. These laws seem to be based on somewhat similar injunctions issued to the children of Israel and recorded in the book of Leviticus. While those laws are entirely defensible on the grounds of the dependence of a nation's strength upon the purity of the family lives of its citizens, it is not necessary to suppose them to have been originally formulated solely in the interest of the progeny. The Mosaic law precludes cohabitation in cases where no actual relationship exists. Familiar statistics regarding relationship of parents of inmates of institutions for defectives and insane will not be cited. In spite of the biased attitude of compilers of such data and the contradictory character of the teaching of the facts accumulated at different times and places, the impression is irresistible that to some degree at least the kinship of the parents is a considerable factor in the production of such abnormalities. Attention must be drawn to the unfairness here, and in regard to acquired characters, of assuming a complete parallelism between our domestic animals and the members of the human family. The developing period of the human infant is so much longer, the impressions bearing upon it so much stronger, and its susceptibilities so infinitely more acute, that the possibilities of environmental modifications, mental and physical, are out of all

Opposition to Inbreeding.

AN ABERDEEN-ANGUS SHOW HERD.

comparison to their import in organisms of lower place. Even a slight inherited predisposition to defective mentality may be fostered by continuous impressions in the home or outside life. The compensating qualities may be so dwarfed by the absence of forces essential to their development that it is quite reasonable to explain the facts as being the result of special circumstances, assisted or not by inherited leaning, which might never be suspected in the presence of counteracting environment.

Some of the injurious effects of inbreeding are set forth in the histories of the breeds. Thomas Bates' Duchess family of Short-horns was bred very closely, probably no less because of a desire to inbreed than because of the inferior character of unrelated stock. A pedigree of one of the members of this family is shown on the opposite page.

Thomas Bates, and Inbreeding.

The intensification of the Duchess blood is further shown in the fact that of the twenty-two bulls used with Duchess cows, The Earl (646) sired five Duchesses; Second Hubback, twelve, and Belvedere, nine. The first was out of Duchess 3d; the second had Duchess blood only through his sire and the third was wholly unrelated though himself inbred. The final condition of the family, with respect to fertility, is shown by Sanders* in the table on page 174 which when summarized shows that up to 1849, from fifty-eight Duchess females of which the first was calved in 1808 and six later than 1841, a total of 110 calves were produced. Of the fifty-eight, twenty-four never produced calves.

*"Short-horn Cattle," pp. 113-114.

INBREEDING AND LINE BREEDING

DUCHESS 55TH.

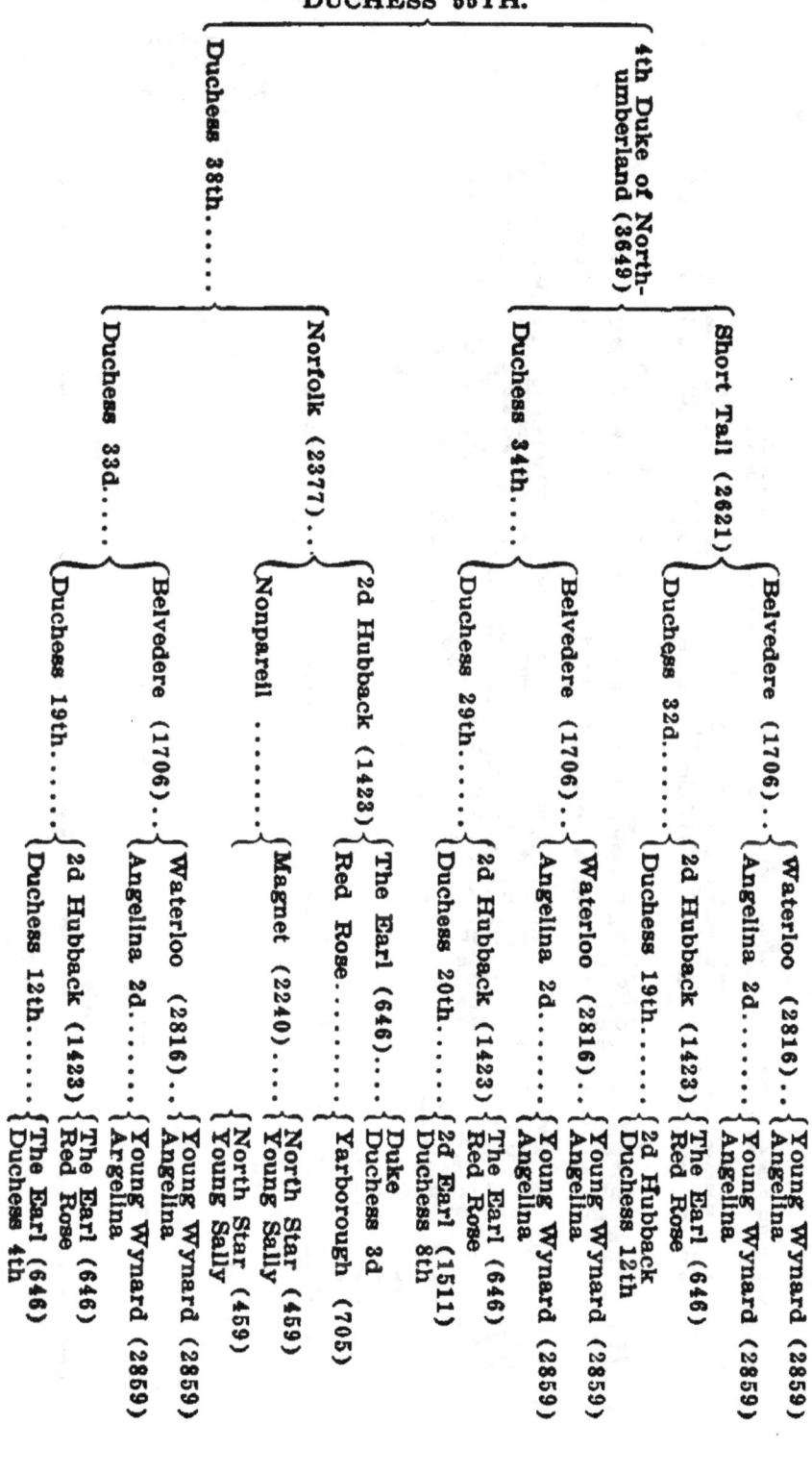

LIST OF PRODUCE OF DUCHESS COWS.

Name		Born	Color	Sire	Dam	Calves produced			
						Males		Females	
						Bulls named	Steers not named	Lived	Died
Duchess	1	1808	r & w	Comet		1		4	
"	2	1812	r & w	Ketton	1	2		2	
"	3	1815	r & w	Ketton	1	4	1	4	
"	4	1816	r & w	Ketton 2d	1	1		3	
"	5	1817	r & w	Ketton 2d			1	1	
"	6	1819	r & w	Ketton 3d	4		1	4	
"	7	1820	r & w	Marske	3				
"	8	1820	r & w	Marske	2	1	1	3	
"	9	1821	r & w	Marske	2		1	2	
"	10	1822	r & w	Cleveland	4				
"	11	1822	r & w	Young Marske	5		1	2	
"	12	1822	red	The Earl	4			1	
"	13	1823		The Earl	9				
"	14	1823	r & w	The Earl	6		1		
"	15	1824		The Earl	8				
"	16	1824	y & w	The Earl	3			1	
"	17	1825	r & w	3d Earl	11				
"	18	1825		2d Hubback	6				
"	19	1825	y & r	2d Hubback	12			5	1
"	20	1825	r & w	2d Earl	8			2	
"	21	1825	r & w	2d Earl	3				
"	22	1826	r & w	2d Hubback	9				
"	23	1826		2d Earl	11				
"	24	1826	r & w	2d Hubback	6				
"	25	1826	r & w	2d Hubback	8				
"	26	1826	r & w	2d Hubback	3	1	1	1	
"	27	1827	r & w	2d Hubback	16				
"	28	1827	r & w	2d Hubback	6				
"	29	1829	r & w	2d Hubback	20			1	
"	30	1830	r & w	2d Hubback	20		3	6	
"	31	1830	r & w	2d Hubback	26				
"	32	1831	r & w	2d Hubback	19	1		1	
"	33	1832	roan	Belvedere	19			1	
"	34	1832	r & w	Belvedere	29	4	1	2	
"	35	1833	red	Gambier	19				
"	36	1834	r & w	Belvedere	19				
"	37	1834	r & w	Belvedere	30	1	2	2	
"	38	1835	roan	Norfolk	33	1		2	
"	39	1835	roan	Belvedere	30				
"	40	1835	roan	Belvedere	19				
"	41	1835	roan	Belvedere	32	2		2	
"	42	1837	roan	Belvedere	30		1		
"	43	1837	red	Belvedere	34	1	1		
"	44	1838	r & w	Short Tail	37				
"	45	1838	r & w	Short Tail	30	1			
"	46	1838	r & w	Short Tail	34				
"	47	1839	red	Short Tail	37				
"	48	1839	r & w	Short Tail	30				
"	49	1839	r & w	Short Tail	30	1		1	
"	50	1839	white	D. of N'th'land	38	1		1	
"	51	1840	roan	Cleveland Lad	41	3	2		1
"	52	1841	r & w	Holkar	38				
"	53	1842	roan	D. of N'th'land	41				
"	54	1844	red	2d Cleveland Lad	49	1		3	
"	55	1844	red	4th Duke of N.	38	1		1	
"	56	1844	r & w	2d Duke	51	1		2	
"	57	1844	roan	2d Cleveland Lad	54				
"	58	1846	red	Lord Barrington	54			1	
						29	18	63	2

Although the Bates Duchesses have been regarded as the striking instance of the results of close breeding, the following facts deduced from Mr. Sanders' table are worthy of study: The later Duchesses were presumably more intensely bred than the earlier ones. In the first half of the period from the birth of Duchess 1st to that of Duchess 58th, or down to 1827, twenty-eight Duchesses were produced. In the last nineteen years thirty Duchesses. The last thirty produced fifty-six calves being no less prolific than the first twenty-eight which produced fifty-four calves. The first twenty-eight Duchesses included twelve barren ones, the last thirty, the same number. The third Duchess had two barren daughters by different sires, so evidently the tendency to barrenness was present in the early days of the family and did not wait until after the closer breeding had been done to show itself. Whatever may be the condition or character that causes barrenness it must have a physical basis and lend itself to intensification through close mating in the same manner as other physical qualities, but it is not shown that inbreeding originated the barrenness in the Bates Duchesses.

Barrenness in Earl Duchesses.

The claim has been quite freely made within recent years that the present type of Poland-China is much less prolific than were earlier representatives of the breed, and that some of the loss is due to the matings of related families to produce the present exceptional individuals. Conditions govern-

Swine Statistics.

ing selection may be considered as sufficient to account in themselves for the result, but inasmuch as it has been stated that the facts are not really as supposed, a discussion is not out of place. Rommel* has compiled statistics from the American and Ohio Poland-China registers comparing the average size of litters from 1892-6 with the average size of all litters registered from 1898-1902 the figures being 7.04 for the earlier time and 7.52 for the latter, thus indicating that there has been no falling off in fecundity. These figures, however, do not of themselves invalidate the claim of decreased fecundity. At best a very small proportion of the animals registered represent the so-called showyard type against which criticisms are principally lodged. It remains to be shown statistically, that the herds of the extreme show type, in which close breeding has been most common, have retained the prolificacy of their more primitive progenitors.

Laboratory Experiments. The result of experiments conducted to furnish data upon the effect of inbreeding, and in which all other influences were carefully excluded, are of more than ordinary interest. Bos,† a German investigator, bred a family of rats for six years without introduction of any new individuals. Young rats were bred back to their parents and females were mated with their full brothers, such being continued for thirty generations. During the first twenty generations there was a slight decrease in prolificacy. The average number of

---
*Bureau of Animal Industry Circular No. 95.
†Reported fully in Morgan's "Experimental Zoology," chapter 12.

young per litter with the initial stock was 7½ and in the twentieth generation 6 21-36. In ten generations succeeding the average size of litters rapidly decreased to one-half the original number, and 41.18 per cent of the pairings were fruitless. An accompanying decrease of 20 per cent in size is also recorded. Breeding of mother to son and daughter to father was less injurious than breeding brothers and sisters. Though of a lower type, rats are as truly mammals as are cattle. While the ultimate results of this experiment are very striking, it is important to observe that the injurious influence on fertility was evidenced only after the twentieth generation of very close matings.

Castle* and his associates have inbred the pomace-fly, (Drosophila ampelophia) for fifty-nine generations. Brothers and sisters were caged together and their offspring selected in the pupa stage for pairing in other separate chambers. Where two pupae developed the same in sex a rearrangement was made to secure the presence in each chamber of male and female from the same parents. Castle's conclusions from this work are as follows:

(1) "Inbreeding probably reduces very slightly the productiveness of Drosophila, but the productiveness may be fully maintained under constant inbreeding (brother with sister) if selection is made from the more productive families.

(2) "In crosses of a race of low productiveness and frequent sterility (race A) with a race of high productiveness (B) it has been found that a female of race A does not have her fecundity increased by mating with

---
*"Proc. Amer. Acad. Arts and Sciences," XLI, 1906.

a male of race B, and conversely, a female of race B does not have her fecundity diminished by a mating with a male of race A. Hence every male not actually sterile furnishes an abundance of functional spermatozoa.

(3) "The cross-breds produced by the mating, B female with A male, are all of high productiveness.

(4) "The cross-breds produced by a mating A female with B male are usually but not always of high productiveness.

(5) "The children of both sorts of cross-breds (see 3 and 4) are some of high productiveness like race B, others of low productiveness like race A.

(6) "Low productiveness is inherited after the manner of a Mendelian recessive character in certain of the crosses made, skipping a generation and then reappearing. In other cases it has failed to reappear in generation F_2, indicating its complete extinction by the cross. In a few cases it has failed to be dominated by high productiveness in generation F_1. In such cases the female parent has always been of race A. Hence low productiveness (or sterility) of the female may be transmitted directly through the egg from mother to daughter, but only indirectly through the sperm, the character skipping a generation.

(7) "A cross between two races, one inbred for thirty or more generations and of low productiveness, the other inbred for less than ten generations and of high productiveness, produced offspring like the latter in productiveness, but not superior to it.

(8) "The same two races crossed after an additional year of inbreeding (about twenty generations) produced offspring superior to either pure race in productiveness."

If it be true that inbreeding contains such possibilities of evil as have in the past been attributed to it, what then was the justification of the practice in the past and

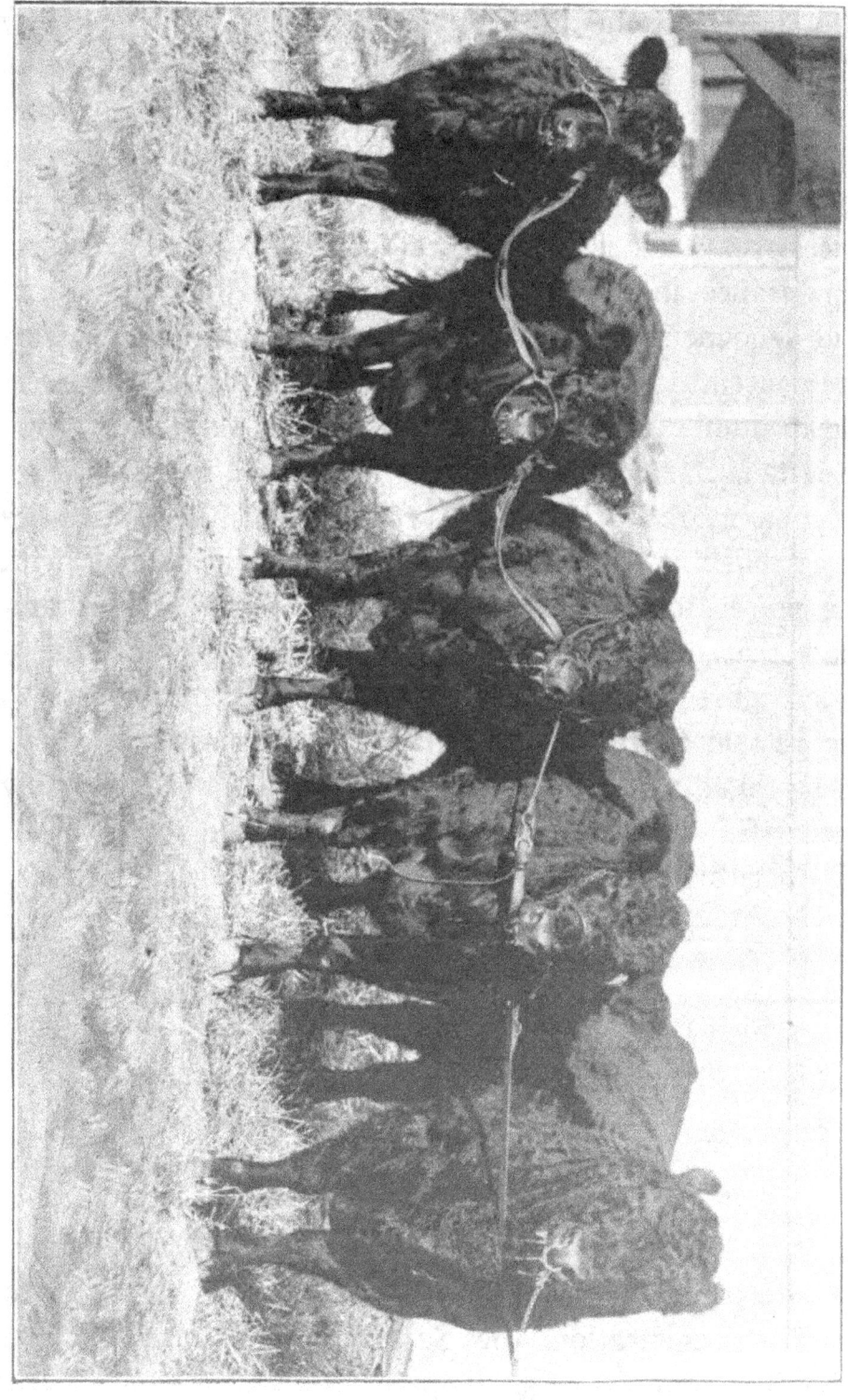

A GALLOWAY SHOW HERD.

to a lesser extent at the present time? It is generally believed that progeny whose parents are related are more prepotent than those resulting from the union of individuals of entirely dissimilar ancestry. This is the natural consequence of the preponderance in inbred stock of the hereditary material and tendencies possessed by the individual with which the concentration begins: not only are units of the germ plasm numerically strongest but their similarity gives a certain number greater power than an equal number from a more varied ancestry. One of the difficulties in establishing a breed is the securing of individuals with the power to transmit the qualities of the various animals that evidence the improvement. In almost every breed inbreeding has been practiced by the founders to secure that fixity of type that entitles a class of animals to be called a breed. The pedigree of Comet, a notably successful sire and sold for $5,000 at the dispersion sale of Charles Colling in 1810, is an interesting study.

Benefits of Inbreeding.

PEDIGREE OF SHORT-HORN BULL COMET.

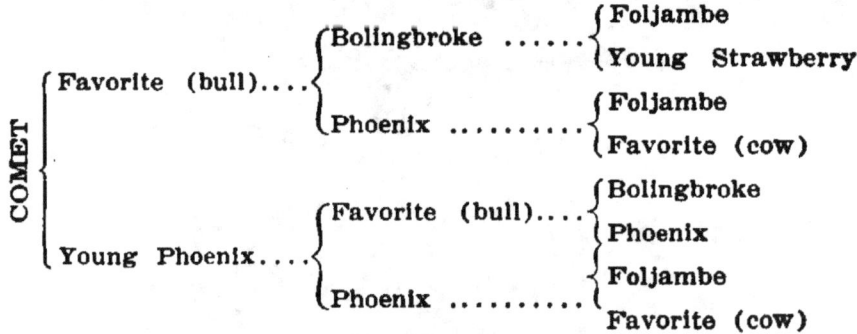

Houseman* states that in 1839, seven years prior to the first registration, the Sovereign blood was in the

---

*"Cattle Breeds and Management," p. 109.

Hereford ranks what Belvedere was in those of Shorthorns, and the name of Hewer parallel with that of Bates.

PEDIGREE OF HEREFORD BULL SOVEREIGN.

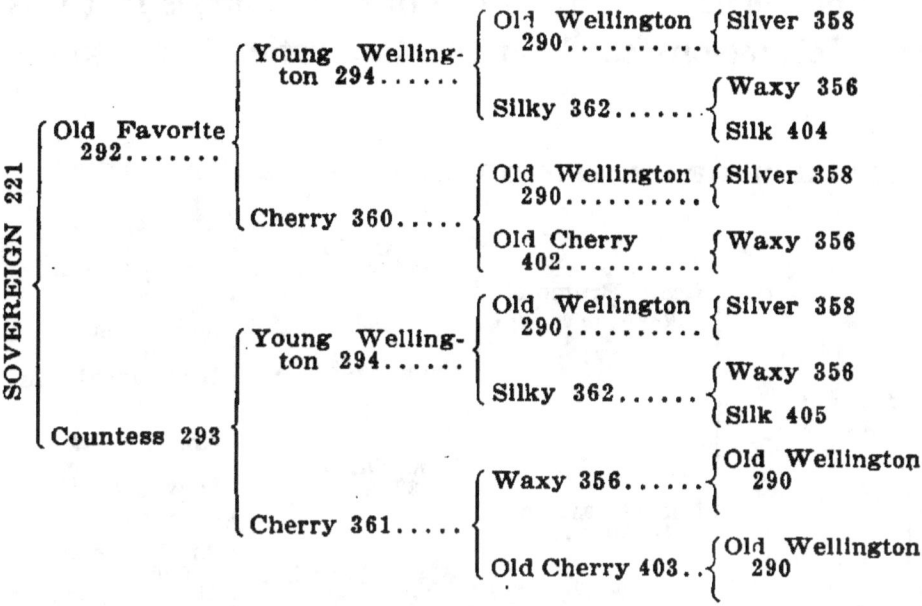

Other notable breeding animals have not been so intensely bred, but among those that stand out now as having marked eras of special progress close breeding is the rule and not the exception. In the fixing of types within breeds similar facts are observable. Amos Cruickshank, whether from choice or through compulsion, is known to have benefited greatly by the practice of inbreeding, though it must also be said that benefits were not all that can be attributed to the practice of that system in his herd.

American Hereford Breeding. The American type of Hereford is eminently more useful in America than is the English type. The progress in the evolution of that breed in this country, while effected by the concerted efforts of a number of breeders who have avoided continued mating of close relationships, has been eminently advanced by Gudgell & Simpson under a system of inbreeding. The breeding of one of their animals that was junior champion female at the International in 1900 is given.

PEDIGREE OF HEREFORD HEIFER MISCHIEF MAKER.

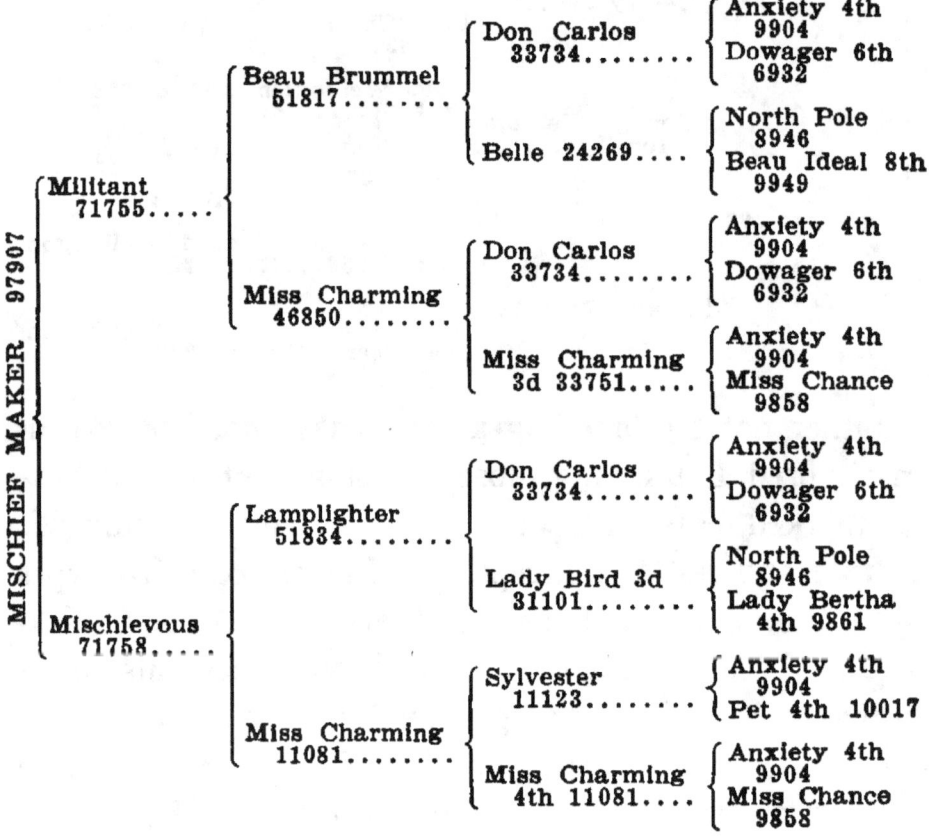

A full sister of Mischief Maker 97907, Miss Caprice 109725, was junior champion female at the 1901 International. The part played by Anxiety 4th in the making of the American Hereford and the prepotency of his descendants is commonly spoken of, but reference is seldom made to his own breeding as a probable explanation of his marked prepotency.

PEDIGREE OF PREPOTENT HEREFORD BULL ANXIETY 4TH.

ANXIETY 4TH 9904
- Anxiety 2238
  - Longhorns 2239
    - Mercury 2241
    - Duchess 2242
  - Helena 2240
    - DeCote 2243
    - Regina 2244
- Gaylass 9905
  - Lognhorns 2239
    - Mercury 2241
    - Duchess 2242
  - Lofty 9906
    - DeCote 2242
    - Fairy 9078

The Gentry Berkshires. Considerable publicity has justly been given the work of N. H. Gentry in his breeding of Berkshires. Though the matings of Mr. Gentry's animals have been by no means so close as those typified in the last tabulations he has not purchased a boar in twenty years, yet his large herd shows no evidence of impairment of size, vigor or fecundity, and has produced an unusually large proportion of prizewinners.

The Principle of Inbreeding. From the two types of instances presented it is evident that inbreeding contains strong possibilities in either direction, and there must be a common principle underlying both sets of occurrences. Major David Castleman has said: "Inbreeding has produced some of the finest successes and some of the most dismal failures. We sometimes use it but feel that in so doing we are playing with fire." One assumption seems justified and that one is sufficient to explain the facts, namely: that the matter of kinship is of itself not the cause of the observed effects commonly attributed to inbreeding, but that the similarity of characters of parents constitutes the seat of the pronounced possibilities of inbreeding. In other words, we may say of the cases that have resulted unfavorably that we should look, not to the kinship of blood but to the kinship of defect. Similarly we may say of the successes of inbreeding, they are attributable, not to the kinship of blood but to the kinship of superiority.

Knowing something of the behavior of the hereditary material, it is possible to explain the intensification of a good or a bad quality common to two parents in the same way as the development and accretion of minor congenital variations was explained in Chapter VI. As with the variations referred to, it must also be remembered that there may be in individuals inherited and recessive defects which come to notice only when intensified and aided by sympathetic matings such as may be looked for in representatives of the same family. Most of the cases of decreased fecundity in farm animals attributed to too close matings are probably due to the intensification of existing weaknesses or hindrances to reproduction.

Inbreeding, Per Se.

It must be recognized however that there is a possibility of a measure of the same result from inbreeding in itself. It is considered by zoologists that in the lowest forms of life in which the union of two individuals is not essential to reproduction, that a greater vigor results from reproduction participated in by two individuals. In some forms individuals reproduce independently for a number of generations and then conjugate. The mixing of material from separate sources appears to add vigor much as is observed in cross-breeding of larger animals. Continued mating of animals retricted to a common descent may then of itself diminish the vitality of the stock. Color is given this idea by the unusual vigor sometimes apparently present in the offspring of two pure-bred parents of different breeds. Such a cross is the extreme opposite of close mating. The same principle obtains in plants propagated by vegetative methods. Prof. Cook* states: "The weakened vitality of old varieties of potatoes or sugar cane may be compared with the gradual weakening of aged trees or of aged men. There is a slackening of the organic energies which can be quickened only by new conjugations." It is possible in animals to so restrict and concentrate the ancestral hereditary material as to render new conjugations imperative. Pronounced injury from inbreeding fully robust animals would only come however from long continuation of the practice as shown in Bos' experiment cited earlier in this chapter. On the other hand it is possible to so dilute and dissipate the heritage of good as to lose what generations of careful breeding have built into the stock. It has been said that in the case of some of our carefully bred families of stock, the paramount question is not how much inbreeding is safe, but rather, how much outbreeding can be permitted?

Risk in Out-breeding.

*Bureau of Plant Industry, Bul. No. 146, page 13.

As stated, however, most of the decreased fecundity of farm animals properly attributed to close matings must be regarded as the result of the intensification of existing tendencies rather than the lack of new conjugations. Such tendencies present in a minor degree may be strengthened into serious defects by close breeding just as may any other bad feature, or as a useful quality may be accentuated and transmitted more strongly by judicious limitation of descent. Sterility that is the outcome of inbreeding must not be regarded as a single characteristic separately transmitted as such. It is doubtless the result of the accumulation or intensification of a number of conditions bearing unfavorably upon reproduction but not previously so strong or so combined as to constitute effective obstacles to breeding. Sterility produced by inbreeding marks the limit, and recovery can be effected only if parents or ancestors still remain to permit a retracing of the course pursued to a point of greater fecundity.

Success or failure with inbreeding is then clearly dependent upon selection. Ability to select necessitates not only well trained powers of observation and good judgment, but also an intimate knowledge of the individuality and ancestry of all the animals in which the breeder is directly or indirectly interested. An examination of the personal qualities and the methods of those men who have successfully practiced inbreeding will reveal in every instance the fact that they were thorough students of the individuality of every one of their animals and in no cases allowed superior

When To Inbreed.

lineage to blind them to the presence or seriousness of an undesirable quality or character. Inbreeding has not been practiced by any successful breeder at the commencement of his operations. The exercise of the abilities of the masters in the art has resulted in their attaining a measure of success that gives them within their own herds animals superior to any that can be purchased. Any present-day breeder who really reaches such a position cannot afford to lightly decide to set his face unalterably against inbreeding.

In comparing modern breeders with those of earlier times one other factor must be regarded. Popularity of strong strains and families within each of the breeds has given members and descendants thereof wide dissemination, and it is difficult to procure within these old breeds good animals so nearly unrelated and free from common tendencies as in earlier days. The fact that Scotch Short-horns were more closely bred than were English Herefords accounts for the present greater aversion of Short-horn breeders to close matings. Some advocates of inbreeding would seem to suggest that selection be based solely on descent. So long as animals are individually adapted to each other and there is no common weakness in their lineage, a degree of common relationship to superior animals is not a detriment but an advantage. Line breeding, as defined on page 168, permits concentration of type by selection from numerous descendants of a good individual and may retain the best features of that individual without the concentration of blood that may cause some minor weakness to be intensified into serious ones.

CHAPTER XVI.

MENDEL'S LAW.

The best farm animals of today are much better suited to present needs than the most popular types of the middle of the nineteenth century. For the most part the present types serve existing demands with greater satisfaction than was experienced by stockmen working with the best animals obtainable for the conditions of some decades ago. It is also not improbable that should conditions of use and rearing of fifty years ago again become operative that some of our prized animals might readily be discarded for the types of their early progenitors. The breeders of each age and each area of country retain those animals best able to do and give what is then and there demanded of them. In some cases breeders have perpetuated and intensified tendencies and characters that seemed to be of advantage in rearing or to give added value when selling, but sometimes selling value has been obtained at the expense of true economy of production as is evidenced in numerous discussions of size and bone, particularly in swine. Low cost of production has also been offset by reduced value as evidenced by market discriminations against animals very large and growthy but coarse and rough.

What changes in animal types the market demands

and farming practices of the future may necessitate can not be foretold. The limit of improvement, or more properly of adaptation to artificial requirements, lies only in the effects of selection for characters that are opposed to growth, health, or the natural exercise of powers of reproduction.

Breeding in the Future.

In discussions up to this point, selection with its essential accompaniments has been prescribed as the basis of progress toward any desired standard. This however assumes the existence somewhere of the component characteristics of the animal it is desired to produce and to multiply. If we have a starting point of even a small variation toward what is desired, the cumulative effects of selection of the fit and the rejection of the unfit will render possible the practical development of everything having an existing basis. Artificial selection accomplishes for artificial needs what natural selection adapts to natural requirements. In each case the law of the "survival of the fittest" obtains. But how should it be if there developed a need for animals with characters not found in any of their kind? "Natural selection may explain the survival of the fittest but it cannot explain the arrival of fittest."*

Beginnings of New Characters.

The first hornless pure Shorthorn of which we have knowledge was not the outcome of generations of gradually declining horn growth. It appeared suddenly without apparent reason and with strength to impress

Mutation.

---
*Thomson, "Heredity," p. 98.

its offspring with its own peculiarity. Such an unusual character appearing without intermediate stages between itself and the usual form is called a mutation and the individual exhibiting the mutation a mutant. True mutants are also referred to as sports or freaks. A horse with a very long and distinctly curly coat would be a mutant if it were certain that it was not a case of reversion. Castle has bred a strain of guinea pigs that uniformly shows four toes on each hind foot, one more than the usual number. The start consisted of one with an imperfectly developed fourth toe. None of this animal's ancestors had been known to show even any rudimentary resemblance to an extra toe.

Polydactylous Guinea Pigs.

Illustrations of animal mutations of practical utility are not easy to suggest. The question of mutations has been studied quite thoroughly in plants where it seems to have greater practical possibilities. The application of the principle has been so sanguinely commended to animal breeders by persons conversant with its use in plants that a review of the matter in that field is of more than passing interest.

There had existed for some time a growing dissatisfaction with the necessity of explaining the origin of all species and varieties of plants by the very slow and gradual operation of natural selection among variations of minor degree. It was felt that some varieties had come into existence more quickly than was probable by this method. De Vries, professor of botany in the University of Amsterdam, had

De Vries' Experiments.

observed distinct changes in plants occurring spontaneously, or at least with no apparent previous tendency in the same direction.* This investigator undertook to secure the double flowered character in the cultivated variety of the corn marigold (Chrysanthemum grandiflorum). This plant averaged twenty-one ray florets to a flower, while the wild form averaged only thirteen. The seeds of six plants were planted separately and in five of the six groups there was a lack of constancy to the twenty-one floret type and consequently only seeds from the sixth group were retained. In 1896 De Vries found in the progeny of one of this group a plant with two secondary heads, having twenty-two florets, the terminal heads still showing twenty-one. Succeeding generations from this plant showed great tendency toward an increased number of florets, forty-eight being reached in 1898 and sixty-six in 1899. Late in the same season three secondary heads were found with florets on the central part of the flower. This was accepted as the arrival of the hoped for mutation. Plants grown from seeds from those heads showed 100 ray florets and 200 in the next generation, a completely double-headed flower. Although the increase in the number of marginal florets was gradual, it may be said that the true double-flowering character appeared spontaneously though it would seem to have been connected with the previous selection.

The cause of the origin of such a mutation is not suggested by the botanist. Manifestly it was not the product of environment. The nearest approach to a satis-

---

*DeVries, "Species and Varieties, their origin by Mutation."

factory understanding lies in the reference of such occurrences to the maze of possibilities in the combination of the chromatic elements of the reproductive cells as discussed in Chapter VI. In more recent years we have received accounts of the production of a new color in plants by the injection of solutions of mineral substances into developing ovules.* Subsequent attempts to secure such results have proved failures. Even were such procedure practical in plants it would not be so in animals. The introduction of any new transmissible element into the hereditary material is inconceivable in view of what is known of that substance. We are again reminded of the limitations of our knowledge of the ultimate source and nature of the chromatin bodies. Procurement of mutations is wholly dependent upon chance and the capacity of the breeder to detect them. The distinction between mutations and variations is really one of degree. It will doubtless be more helpful to think of future modifications of animal form as having their beginning in minor variations occurring without design and offering opportunities to those best fitted to recognize and utilize them. Any new feature promising value, no matter how little developed, when favored by an encouraging environment and the most careful selection, may in course of time be brought up to a useful degree. However, should mutations of pronounced utility present themselves they may be utilized even if found in inferior stock, as has been done in forming our single standard breeds of polled cattle.

Cause of Mutation.

---
*McDougal: "Science," Jan. 24, 1908.

A SHROPSHIRE SHOW FLOCK.

Each of the breeds of live stock, even of those kept for the same purpose, is characterized by special features of excellence. Of course breeders of each of the competing breeds, while retaining the admitted superiorities of their own stock, try also to secure as much as possible of the pronounced good features of their rivals. Not infrequently the crossing of established breeds is resorted to in the hope of combining the good features on both sides. Bearing in mind that every quality, desirable or undesirable, is represented in the hereditary material, regard must be had for the fact that in a cross there is no virtue that can obscure the weaknesses. These must be expected with the rest and are just as likely to be contributed from both sides as are the good traits. The crossing of breeds is sometimes favored as a means of securing new variations and new forms. The mixture of hereditary material from two dissimilar sources, each of which has been rendered pure to its usual properties and at no point allowed to receive any vestige of a taint from the other, may yield unusual results. Such procedure is quite practical among plants where many random trials are practicable even if only one in thousands yields anything of promise, and it is by crossing that Mr. Burbank has secured some of his more useful plant creations.

Cross Breeding for New Characters.

The making of a cross cannot be expected to originate any new character; yet the breaking up of forces among old tendencies may so balance and engage one another as to give opportunity for previously dormant and restrained possibilities to evidence themselves.

Crosses sometimes show likenesses to very remote parents, which likenesses really constitute reversions as exemplified in the case of a cross-bred (Hereford-Shorthorn) bull bred to a pure Angus cow. The offspring was creamy white in color with black muzzle and black hair in the ears, features of the very early native British cattle.* In the main, however, the most that can be looked for in crossing is a fortuitous combination of existing characters such as is evidenced in the offspring of a Hereford and Angus cross which exhibits the polled head and black body together with the white face. The Oxford breed of sheep was made by selections from Hampshire and Cotswold crosses, and is the only breed of consequence that has resulted from crossing old established breeds.

The demonstration of the actual occurrence of mutations as sudden and considerable departures from the usual order of things encouraged much hopeful speculation as to changes of magnitude the breeders of the coming years might effect. It may be repeated that the hopes of accomplishment among animals was based on a somewhat over-drawn analogy between the plant and animal kingdoms, especially in economic aspects. Even more stirring than the mutation theory was the announcement at about the same time of the discovery of Mendel's law.

Mendel. The discovery of the operation of this law or principle was announced almost simultaneously in 1900 by De Vries, Correns and Tschermak. It was soon learned that similar work had been done and similar conclusions

---
*Thomson, "Heredity," p. 378.

published in 1865 by Gregor Mendel, a monk in an Austrian monastery at Brünn. In view of the priority of Mendel's work and announcements his name is always used to designate this interesting phenomenon. Mendel's law is of special interest in conjunction with mutations as suggesting practical procedures in their perpetuation. It also throws some light on the behavior in transmission of existing characters, and is therefore worthy of careful study.

Mendel was also a botanist, and the experiments which led up to his discovery were conducted with the common sweet pea. The stockman's interest in Mendelism may be better discussed after the law has been explained in its applications to the plants with which its discoverer worked. The law as set forth by Mendel does not lend itself to terse statements and it will therefore be more satisfactory to outline the experiments in their natural order. In 1857 Mendel planned and inaugurated the experiments which at the end of eight years justified such important conclusions.* Two of the varieties of peas selected represented extremes in regard to length of stems, the plants of one uniformly having stems measuring from 6 to 7 feet in length, while in the other the range was limited with equal uniformity to between 9 and 18 inches. These two varieties were crossed, and when the resulting seeds were planted the following season it was found that all of the plants had stems fully equal in length to those of the longer stemmed parents. These cross-bred plants

*Bateson: "Mendel's Principles of Heredity."

while not hybrids in the true sense are erroneously so designated for convenience. The chromatin representing the short stem was of course present in the hybrid plants but had not asserted itself, and the short stem character in this case is therefore termed recessive and the long-stem character dominant. Those hybrid long-stemmed plants were pollinated exclusively by pollen from plants of their own group. In the third season there were reared as the progeny of hybrid parents on both sides 1,064 plants. Of this number about one-fourth, or 277, had the short-stem character of their grandparent which had been recessive in the parent, the remaining three-fourths, or 787, all had long stems. This was a very suggestive fact and Mendel proceeded to investigate the breeding qualities of these two groups. The flowers of each plant were fertilized exclusively by pollen from plants of their own kind. Six other similar experiments were carried on simultaneously with such characters as position of flowers, form of pods, and form of seeds.

As it was impracticable to breed from every plant reared, 100 plants were selected from each of the two groups preserved from the long and short-stemmed hybrid lot and their seeds were planted the following season. The progeny of the short-stem plants exhibited short stems exclusively. Of the 100 representative long-stemmed plants it was found that 28 produced only long stems, while 72 produced some of each kind. This proportion is approximately three to one, but applying the actual figures to the whole number it shows that the 1,064 plants really comprised 277 capable of producing only short stems; 567 capable of producing either, and

220 capable of producing only long stems. Each of the two smaller groups continued to reproduce their own kind exclusively through succeeding generations as long as bred within their own groups. Further work with the larger and unstable group, however, showed that it behaved exactly as did the original hybrids giving off one-fourth of its number to produce only long stems, another fourth for short stems, and the half of hybrid character to again break up into the three kinds.

First year.—Long and short stems crossed.

Second year.—Hybrid plants raised from seed produced the first year; allowed to fertilize each other.

Third year.—Seeds produced the second year planted and 1,064 plants reared; of these, 277 had short stems and 787 long stems; plants of each group fertilized by their own kind.

Fourth year.—Two groups of the third year were tested; short-stem group found to produce its own kind exclusively; of the larger group 220 were found to be pure to the long type, while the seeds of the other 567 produced both long and short stems.

Fifth year.—Seeds planted from offspring of each of three groups revealed in previous season. The offspring of the pure long and pure short groups bred true. The offspring of the 567 plants producing mixed progeny, breaking up into 142 short-stem plants and 425 with long stems. On further test those two groups prove to behave exactly like the two groups of the third year.

Knowing the breeding records of the various groups, the divisions of the original number may be arranged so as to show their relationship.

MENDEL'S LAW

It must be borne in mind that while the plants pure to the recessive short-stem character in generation two were separable as soon as they appeared, the 220 with no tendency to short stems could be separated from the larger groups remaining mixed only by examination of the plants grown from their seed. In all instances plants were fertilized by others of their kind.

FOUR GENERATIONS OF CROSS-BRED PEAS.

Generation 1.	Generation 2.	Generation 3.	Generation 4.
	277 pure short stems inter-bred produce short stems exclusively.	pure short	pure short.
Hybrids: All long stems: inter-bred:	567 hybrids: inter-bred: produce both long and short.	142 pure short	pure short.
		283 hybrids	71 pure short. 141 hybrids. 71 pure long.
		142 pure long	pure long.
	220 pure long: inter-bred produce long only.	pure long	pure long.

Mendel's* Law.

The discovery thus made was that hybrid parents produce offspring of which one-half are again hybrid while one-quarter are pure to each of the original parent forms. The figures 277, 567 and 220 are only approximately in the proportion of 1, 2 and 1, but the entire numbers of offspring could not be tested and the figures used represent the rather unfair proportions derived from the actual test groups. It is not claimed that the proportions will occur with exactness except in very large numbers, although the summary of all Mendel's tests shows a very close adherence to the set proportions. From a single individual possessing a desired character or mutation that obeys Men-

---

*A translation of Mendel's original papers appears in Bateson's "Mendel's Principles of Heredity."

del's law it would be possible in time to procure a large number of others equally strong in the same character. As will readily be seen, however, progress would be much more rapid in dealing with characters that are recessive since they may be selected as soon as found in the second generation from the cross without awaiting the breeding test that is necessary to segregate pure dominants from hybrids with the dominant character. Also in practice the hybrids can be rebred to the parent and thus one-half the offspring will possess the parental character. Many characters are not Mendelian and so do not remain distinct but mix with their opposites. Mendelian or non-Mendelian characters can only be determined as such by test.

At first thought the occurrence of the Mendelian proportions seems to be entirely out of line with all ordinary procedures of nature; however, a very plausible explanation is at hand. In the experiment referred to for the purpose of explaining the law it was seen that short-stemmed plants bred to long-stemmed ones produced hybrids, all with long stems. The hereditary element for the short stem was restrained from showing itself but must necessarily have been present in the germ cell material of the hybrid. It manifests itself in the production of 25 per cent of the offspring of the hybrids that contain no long-stem material, as is shown by the fact that they and all descendants remain true to the short-stem character. The hereditary material contained in the reproductive organs of a hybrid individual must contain elements for both characters;

How Mendelian Proportions Occur.

whether each chromosome received from the long-stemmed parent contains the element for that character or whether it resides only in a single chromosome, our knowledge of those bodies will not permit us to conjecture. However that may be, the occurrences strongly suggest that each germ cell produced by a hybrid parent represents only one of the characters. If this be true, and inasmuch as the stock of hereditary material is equally supplied with both kinds, then the number of germ cells of one kind will be the same as for the other. Speaking only in regard to this single character, it will now be seen that only two kinds of ova can be produced, those with the element for short stems and those for long stems. The same is true of the spermatozoa. Any ovum that is fertilized is as likely to be of one kind as of the other. Any particular spermatozoon sharing in fertilization has also equal probabilities of having the long or the short-stem element.

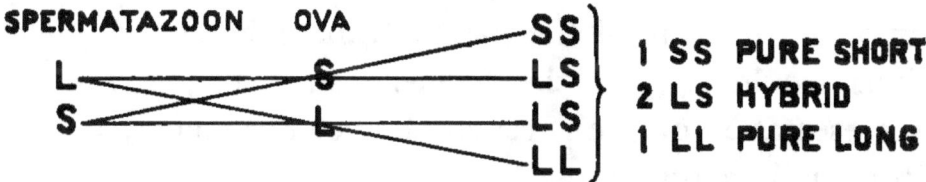

FIG. 12—OCCURRENCE OF MENDELIAN PROPORTIONS.

In mating a female and a male hybrid four cases may arise, as shown in the diagram, Fig. 12: Case 1.—The ovum S is the one presented and is fertilized by the spermatozoon S; the progeny SS, can produce only short stems so long as bred to others like itself. Case 2.—It is equally likely that the ovum S would be fertilized by the spermatozoon L, giving LS, a hybrid progeny which like his hybrid parents might produce either kind of germ cells. Case 3.—It is equally likely that the ovum presented would be of the L type; if so and fertilized by spermatozoon S another hybrid offspring would result. Case 4.—An L ovum joined by an L spermatozoon would produce an offspring pure to the long type. The chances for SS, purity of short stems, are the same as for the opposite, while the probability of LS, the hybrid form, is twice as great as for either of the pure forms. It is therefore not surprising that in a considerable number of cases the proportions of 1, 2 and 1 should appear. On this basis the cause of the Mendelian proportions is apparent.

Purity of Gametes.
Botanists usually speak of germ cells as gametes. This explanation of Mendelism assumes the purity of the gametes to a single character. Whether the gametes are actually pure in one character or whether they contain a predominating amount from one of the parents cannot be stated. The facts would suggest the former and that perhaps the representation of one parental character is conveyed only in a single chromosome.*

Mendelism in Animals.
Before discussing Mendelism among animals it will be of value to gain a clear idea of unit characters. It is fully established that some animal characters follow Mendel's law in transmission. Long-haired and short-haired guinea pigs mated in the experiments of Professor Castle† gave progeny with

---
*Castle: "Carnegie Institution," XLI, 1906.
†"Proceedings American Breeders' Association," Vol. I.

the short-haired character dominant. The progeny of these hybrids consisted of twelve short-haired guinea pigs to four with long hair, the recessive character. This is the expected Mendelian proportions, 3 to 1, for the first generation from hybrid stock. The twelve when tested were found to contain four that bred true to shortness of hair and eight still producing mixed progeny. The same results were apparent in crossing albinos and colored guinea pigs in the same experiment, the colored coat character being dominant over the albino condition. The two characters were studied in the same animals. Starting with long-haired albinos and short-haired pigmented or colored animals, sixteen guinea pigs were reared from the hybrid stock. These sixteen comprised four kinds, as follows:

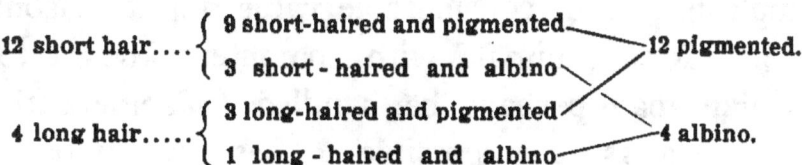

Each of the groups of twelve shown on left and right, on further test was found to be made up of four pures and eight hybrids.

The results of this experiment illustrate the meaning of the term unit characters. It is apparent that there was no relation between the length of hair and its color; each was transmitted entirely independent of the other. The length of hair is therefore one unit character, while the color of hair is another.

Unit Characters.

This experiment also illustrates the practical bearing

Application of Mendel's Law.

of another point. If long coats and albinism were infrequently found and were preferred it would be possible to increase the number of individuals pure to either character to one-fourth the whole number that the second generation might comprise. These two characters being recessive could easily be recognized in the individuals pure to them. Should it be the dominants that were sought for, a breeding test would be necessary to isolate them from the hybrid forms. Among plants the rearing and preservation of such numbers of individuals may be attempted even in a commercial enterprise. Among larger animals however such practice would be difficult even in large experimental work; though should it be possible, accomplishments of considerable value might reasonably be hoped for. Animals with characters known to be Mendelian may perhaps be handled to some satisfaction so long as it is possible to base selections completely upon the one specific character. Conditions may be imagined under which such a course would be practicable but are not likely to arise.

If it were sought to combine in the same animal two unit characters that were also Mendelian, one in sixteen of the second generation could be expected with that combination as is seen in the case of the long-haired albino guinea pig representing the fourth group shown on page 203. If a combination of dominant unit characters were desired the detection of the individual of that class would of course be more difficult.

Many features of comb and plumage of poultry have

A HAMPSHIRE FIELD FLOCK.

been shown to be inherited in accordance with Mendel's law.* The waltzing habit in a variety of fancy mice has been shown to be a Mendelian recessive. This was found to be associated with a peculiar lack of development of the semi-circular canal of the ear which supposedly accounts for the waltzing movement. It has been suggested that the pacing gait in horses may be a Mendelian unit character. So far however this gait has not been found to be uniformly associated with any physical peculiarity and the manner of its appearance does not indicate that it can be regarded as Mendelian. Professor Spillman has presented figures which seem to show that the hornless character when appearing in horned breeds of cattle is a Mendelian recessive. Breeders of blue Andalusian fowls experience great difficulty in breeding the desired color. Birds of the desired colors when mated produce numbers of black, blue and white in proportions of 1, 2 and 1. The blue is regarded as the hybrid though how it splits up into the two colors while they appear to blend in development is not clear. Data has been presented strongly supporting the idea that red and white colors in Short-horn cattle are Mendelian characters.† If such be the case, however, neither can be regarded as dominant since the roan hybrid exhibits both colors.

Limitations of Mendelism. All the animal characteristics that have been mentioned as illustrating unit characters following Mendel's law are external ones and of only secondary importance in breeding. Where any

---

*"Inheritance in Poultry," C. B. Davenport.
†"Breeder's Gazette," July 15, 1908.

of them are desired and can be selected for without interfering with more essential features some advantage may be gained. Selection based solely on a single external character is entirely impracticable in regular breeding. Valuable mutations have been perpetuated and added to existing types in a comparatively short time by breeders who had no acquaintance with Mendel's work, as is shown in our double-standard Polled Durhams. It does not seem that the time required for perfection of the Polled Durhams was any greater than it would have been if the breeders had sought to utilize a knowledge of Mendelism while regarding a polled head as only one of many essentials. Some advantage would have been afforded by familiarity with the law in the certainty regarding what the results would be.

While length and color of hair are separate unit characters in guinea pigs, it would seem the length and fineness of wool fibres may be unit and *Non-Mendelian* Mendelian characters. This, however, is not true of the Cheviot-Leicester cross and, as was stated, the Oxford Down represents and transmits a blend of two distinct types. Probably one of the most interesting things about Mendelism is the evidence that the animal body represents a vast number of unit characters which may or may not be transmitted in definite proportions. Investigators have not attempted to determine unit characters of the body, though it has been suggested that the width of hindquarters of beef cattle is a unit character.*
It has occurred to the author that the size of the Percheron

---

*"Proceedings American Breeders' Association," Vol. 4, page 325.

foot may be transmitted as a dominant unit character and also that peculiar "typey" head of the Berkshire may be a Mendelian unit character. No data have been collected on these points. Additions to our knowledge of what are unit characters of the body are not likely to come through planned experiments. They are more likely to come through retrospective study of records of regular breeders who record and preserve full data regarding the individual peculiarities of each animal reared or used in breeding. Mendelism also suggests the value of careful inquiry regarding individuality of ancestors, to guard against the existence of undesirable characters that cannot be recognized because recessive but which might reappear.

Need of Breeders' Records.

Although the results of Mendel's work have been freely spoken of as promising to revolutionize practical breeding it now seems that the matter is still chiefly of scientific interest. The breeder may aid the scientist by preservation of records that will facilitate this study and in turn the scientist may then be able to make suggestions of possible application in breeding.

CHAPTER XVII.

BREED RELATIONS.

A breed of live stock is not of itself an end but a means to an end. That end is the yielding of a product of maximum value at a minimum cost. The founders of our existing breeds did not set out with a purpose of establishing breeds of animals.* Who would be more astonished to learn the number of white faced cattle in western America than would John Hewer? The makers of our breeds aimed in each case to raise such animals as would be more profitable under the conditions of rearing, feeding and selling, which prevailed in their respective localities. It was when their stock had demonstrated extraordinary merit in the eyes of the buyers that the increase was in demand for use as breeders on other farms and in other localities where the stockmen were not content to repeat for themselves the slow and studied steps to improvement. With the broadening demand in their home countries and the active foreign trade, the registration of pedigrees with its attendant features of good and evil became a necessity.

The Place of Breeds.

We are often prone to measure the ability of our present breeders by the resemblance of their stock to that

---
*See Darwin. "Animals and Plants Under Domestication," Ed. 2, Vol I, page 96.

of earlier masters of renown to whose work the lapse of time has given a clear perspective view. That standard is a false one. Those breeders achieved renown through seeing the peculiar needs of their times and localities. They produced such types and fixed such characters as progressing agriculture and evolving markets demanded, and the only fair way to appraise our present-day types is by considering the degree to which they satisfy the market and, what is of equal importance, the cost of their rearing and finishing in those sections to which their peculiar and distinctive features best adapt them. In many cases the most useful types of to-day are radically different from those of two decades ago. Evolution is continuous, both in our markets and in our systems of cropping and feeding. Since we cannot see far ahead we are safest in setting our standards fully abreast of the times and being thereby best prepared to make such modifications as the future may necessitate.

Evolution of Types.

While all the users of draft horses require many fundamental points in common their varying classes of service and their dissimilar ideas of indications of efficiency and durability give outlets for the array of good types represented in several breeds. While all dairy sections aim at the economical production of milk solids, the demand in some cases makes fat the paramount constituent, solids not fat in others, and in other instances natural color is emphasized. Some soils produce digestible animal nutrients

Need for Numerous Breeds.

more satisfactorily in coarse and bulky feeds than in concentrated forms. The first section necessitates the use of large strong cows developed for and adapted to such conditions. Thus even in milk production a variety of farm conditions and of consumer's demands require several sorts of animals. The same is true of beef cattle, sheep and swine. When we except wool, the demands the markets make on our meat-producing breeds are less varied than with dairy stock, though the range of conditions governing rearing and finishing is much wider.

In European countries there is much of uniformity in the stock of a particular locality. On each of the many types of soils subject to a common climate the farmers appear to have found what marketable articles they can produce especially well. In some instances it is early lambs, in others mature muttons; or again it may be baby beef, finished bullocks, or store or feeding stock. General agreement as to the object and method of the stockman has also resulted in agreement as to the type and breed of animal best adapted to the special requirements.

Distribution of Breeds.

Fortunately or unfortunately our American importers seem to have been less unanimous than the Britons themselves in their estimates of the proper spheres for the various breeds. One has pinned his faith and his reputation to one breed for his native locality, while his neighbor is equally assured that another breed is peculiarly adapted to the same channel of usefulness. Recognizing the superiority of any breed over no breed, all have been eagerly received and opportunity and unstudied impressions have had more to do with choice of breeds than has conviction derived from acquaintance with the objects and requirements by which the breed-makers were guided in the selection of material for their accomplishment. Thus we have owners of different breeds of the same class of stock working toward a common end, and different owners of the same breed building on divergent lines. That individuality and strain are of greater importance than breed is nowhere so true as in America.

It seems reasonable to suppose that with our country becoming more fully occupied and our agricultural practices assuming a more stable aspect the time will come when the live stock in each community will be less varied within small areas that we now find it. On most of the farms located in the area of a single type of soil, and tributary to a few principal markets, we may expect that all animals entering into commerce will leave the farm at approximately the same age and in much the same condition. If such should be the case the methods of rearing and feeding within that vicinity will have more in common. Common interests will then have as their result the maintenance of a common type of animals especially well fitted to satisfy their particular markets and possessed of the qualities to render them especially profitable to their raisers. For the great variety of conditions found in so large a country as the United States many types are needed. Here capacity to mature quickly on forced feeding is paramount; there, a disposition to graze and be less dependent on feeding is more desirable, while again there will doubtless always be some sections

PERCHERON SIX-HORSE TEAM.

that will not finish meat-producing stock but wish to raise what will breed most freely and satisfy distant farmers who are fatteners and not growers of stock.

The greatest uniformity in farming and feeding practices may give each breed full possession of its special area as it has in England where the breeds were developed to answer recognized needs. Should there be such a condition it might greatly reduce some items of expense of production; instead of neighborly rivalry and disagreement as to the best means of reaching a common purpose such as now obtains through the variety of types and breeds in the same locality, there might be partnerships and co-operation in the purchase and use of sires and nearer markets for surplus breeding stock. Of recent years a good deal has been done in Wisconsin and Michigan in the organization of community breeding societies. A number of breeders in a community pledge themselves to adopt the same breed of cattle.

Community Breeding.

They thereby insure the production of a sufficiently large number of surplus stock to attract buyers to them when making purchases. Even if members of the society are breeding only for the market the advantages and economies resulting from the co-operative purchase and use of the best sires the breed affords amply justify the maintenance of the society.

Our state and national and international live stock fairs and expositions are our greatest educational factors in animal husbandry. Their first service is to fully acquaint people with the common requirements of all acceptable market animals. Of necessity the making of awards must relate more to merit of the finished product than to evidences of value in the making of large and economical gains. This is true even in the breeding classes, because the former is more easily discernible than the latter and its requirements are largely common to animals produced in all sections. Then too, should a judge attempt to exercise his opinion to indicate which class of animals is most profitable to their raisers, he must have in mind some particular set of conditions and system of rearing. In a local show a well qualified judge might properly follow such a course; but in a ring of exhibits gathered from a half-dozen or more states he would do justice to only a fraction of the exhibitors and on-lookers.

Value of Shows.

Basis of Awards.

The value of exhibiting for purposes of education and stimulation is especially shown by our breeders of trotting horses and dairy cattle. Breeders of these classes of stock recognize clearly that actual tests of merit in comparison with unvarying standards—the watch and the scales—while not the sole guides are nevertheless much more useful than comparison with ideals of one, two or three men, be they ever so capable and experienced. In spite of the fact that showring awards in these breeds are of only secondary importance to the breeders in their selections, the promotion of interest and study resulting from the shows give such events a position of greatest importance among the many methods for the advancement of animal husbandry. The data afforded breeders of speed horses and dairy stock are chiefly valuable as showing which individuals and combination of strains breed high efficiency with the greatest regularity. Records alone cannot dictate how animals should be mated. Breeders of draft horses and of meat-producing stock recognize the desirability of some form of advanced registry or means of measuring and recording the merit of individuals and their offspring. Manifestly no test can be conducted with meat animals that will permit of their subsequent use as breeders. It is conceivable that tests may be provided to show the amount and cost of gains of breeding cattle, sheep or swine, but such tests can at best be only suggestive. For those who will use them properly and not over-estimate them, year books based upon show awards such as are now being prepared by some breed associations are of assistance to breeders in studying the achievements of representatives of various blood lines.

Advanced Registration.

It is not to be expected that a very large percentage of our farm animals will ever be registered. Indeed if all animals now bearing recorded pedigrees were actually capable of effecting improvement upon the best of the unregistered flocks and herds, the breeders would be more highly regarded and more freely patronized than they now are. Animals with recorded pedigrees are manifestly intended by their owners for breeding purposes rather than for immediate commercial uses. Such pure-bred stocks are valuable and

Registration in The Future.

can be maintained only as they can and do exert an elevating influence upon herds of which the increase is designed for market rather than for further breeding use. In view of these apparent relations it would seem altogether natural that we should expect in the future, not so much a large extension of the practice of registration as the attainment of a much higher degree of actual merit all through the registered animals of herds and flocks designed to serve as sources of improving material. With the realization of such a condition the business of rearing commercial stock will doubtless be more sharply differentiated from the breeding business than it now is and the patronage of the breeders will be more general as well as more discriminating.

The production of market stock by the crossing of distinct breeds is not uncommon in England and some sections of America may find justification for making pure crosses.

Cross-Breeding.

It cannot be denied that for the most part the cross-breeding now practiced results mainly in loss and disappointment. This is due largely to indiscriminate and purposeless crossing. The basis of most of the unstudied practice of crossing breeds is the hope of combining desirable features of both parents while at the same time excluding the less valuable qualities. Crossing is fully as likely to result in a combination of the objectionable features of both sides to the exclusion of the good. It is a very suggestive fact that in only one instance, namely, the Oxford Down sheep, has a new and useful breed resulted from a union of two other distinct breeds. The uncertainties attaching to the offspring of cross-bred parents may doubtless be explained by the principle of Mendel's law as discussed in Chapter XVI.

In some of our more numerous breeds we have two or more distinct types bred to different standards and for different special uses. Such types while similar in many external characteristics have very dissimilar hereditary tendencies, and while recorded in the same book have had their rise and whole course of ancestry with little or nothing in common. Such types within breeds can be mated only with the results attaching to the mating of distinct breeds. There are however some material advantages to be derived from the intelligent crossing of breeds. Cross-bred animals often have a vigor and a robustness greater than characterized either parent. This enhanced vigor gives greater and more rapid growth and therefore permits a considerable economy in the production of a market carcass. The value and uniformity of particular crosses can be ascertained only by tests.

Crossing Types.

Other advantages are found in the greater prolificacy and better nursing qualities of the females of some breeds which when crossed give progeny of satisfactory character and at a lower cost than is common in the stock of the sire. In order to secure the benefits of cross-breeding without the possible losses it is necessary to proceed only in the light of careful experience and to keep the parent stock pure. The demand for good pure-bred females to be used as dams of

Pure Breeds for Crossing.

cross-bred market animals may in the future absorb considerable numbers of the breeders' surplus females. Lowered vitality in dams of market stock may as often be remedied by carefully chosen males of the same breed as by those of a different breed.

It is impossible to foresee the ultimate development of our domestic animals. It sometimes seems that perfection is well-nigh attained. The foregoing discussions have shown, however, that perfection of animal form and qualities is a relative matter. The most nearly perfect animal is the one that best performs what is desired. Inasmuch as what is desired will continue to change as it has always done, the standard is a shifting one, and since it is impossible to know what will be required of future races of animals, conjecture as to their character is of no avail. Many individual animals do certainly approach very close to present day perfection. Such successes of the breeder's art give encouragement and point the way for owners of mediocre stock, and through their own kindred and offspring the good animals facilitate the elevation of the general stock to a higher level. It seems that the most useful work for the breeder now lies more in the production of sufficient numbers of representatives of best existing types rather than in the modification of the stamp of our champions.

Limits of Improvement. The idea sometimes finds expression that improvement can be carried too far for real utility, that the types held at higher levels lead artificial lives and lose the robustness and regularity of reproduction so essential in commercial stock. Cases can be cited to show that herds and flocks bred closely to their breeders' ideals have become too delicate and too low in rate of reproduction to commend them to a place in even the best farm practice. Such so-called improvement has been best illustrated in the passing type of light-boned and under-sized swine, though similar conditions have been evidenced in other classes of stock and give rise to the idea of over-improvement. The term mis-improvement is more nearly correct. Such stocks may have made steady progress toward the standards of their breeders, but it is evident that the ideal of those breeders was not a practical animal and cannot therefore be considered as strictly high-class from an agricultural standpoint. Perfection is too likely to be regarded as residing in the animal form alone. Real perfection embraces conformation and all that is apparent to the eye, but no less it includes those qualities of adaptability to the animal's real work which may mean either spirit and ease of movement, good feeding and digestive powers, or grazing disposition. And in all cases it must include that vitality and constitutional vigor without which there can be neither economy of increase nor certainty of reproduction.

Effects of Injudicious Breeding. In the grading up of native animals it has seemed desirable to eliminate some of the qualities intensified by nature's careful selection for the sole purpose of fecundity and adaptability to natural surroundings. In so doing chief emphasis has been laid upon those features and qualities chiefly demanded by artificial environment, and only after the loss of much of the native regularity of reproduction and foraging capacity have those qualities been fully appreciated. Artificial selection can produce and perpetuate the animals needed for artificial demand with just as much safety and certainty as natural selection has operated to fulfill nature's requirements. The task is a more complex one however, and can only be met by making our selections on a very broad basis, one that includes all useful natural and produced features. There is full assurance that a selection that recognizes fertility and constitutional vigor can improve and intensify those qualities just as surely as selection has controlled the set of the pastern or the turn of the ear.

Breeding for Vigor and Prolificacy.

Achievement of the object of such selection calls for indefinite time, pains and experience. The future of our breed rests with our breeders as a body. No one works for himself alone; each one either retards or accelerates the rate of progress toward the highest usefulness of man's servant companions, the farm animals.

CHAPTER XVIII.

BREEDERS' ASSOCIATIONS.

Origin of Registration.

The aim of the various associations of breeders is to do by combined effort what unorganized individuals could only do with difficulty if at all. The value of combined effort first became manifest when trade became general in highly bred animals. The value of a strong line of ancestry as a reinforcement of individuality in breeding stock was recognized by the patrons of Bakewell and succeeding breeders. So long as the buyer was dealing with Bakewell or the Collings he knew that his purchase was the product of several generations of selection by the person by whom the ancestry was described, and nothing more was necessary. With a rapidly increasing number of breeders and the necessity of dealing with comparative strangers, whose herds and whose careers as breeders were not familiar, it became necessary to have something more than a mere verbal statement of the ancestry or pedigree. Especially was the verbal transfer of pedigree unsatisfactory when animals were bought from persons owning large herds composed mainly of individuals reared by other breeders. The danger of confusion and intentional or unintentional misrepresentation of pedigrees was considerable.

It was with an aim to remove such difficulties that confronted early Short-horn breeders that George Coates, acting on his own initiative, collected the pedigrees of Short-horns of note up to the time he issued his first volume in 1822. It was not until 1876 that the British breeders, organized as the Short-horn Society of Great Britain, took charge of the preparation and publication of pedigree records. The following from Coates' first volume is of interest: "As it must be the interest and wish of every breeder to be enabled to breed with the greatest possible accuracy as to pedigrees and so forth, therefore, to assist the author in correcting errors in this work he takes the liberty of recommending every breeder to have one of his works interleaved with plain paper, and on it to correct any errors he may discover, or make fresh entries as his stock increases; all of which entries (or a copy of them), being sent by every breeder to the author immediately (or at all events prior to the compilation of the next edition of this work), will be an incalculable benefit to the breeders at large and the author in particular."

Short-horn affairs had assumed considerable proportion in America before the time the British breeders took control of their herd book. From 1846 to 1882 Lewis F. Allen performed for the American breeders a service similar to that of Coates in England. Breeders in Ohio and Kentucky also published rival books and it was not until 1882 that the three enterprises were combined under the auspices of the American Short-horn Breeders' Association. The advantages of a single association for each breed are too manifest to need mention. However among those interested in some of our breeds of horses, sheep and swine there is not even yet sufficient harmony of effort and method to prevent the existence of several rival associations organized for exactly similar purposes. Unanimity of effort is desirable because it simplifies registration and exchange of animals of the breed, and also because organized effort can care for important matters better than the most earnest of scattered endeavors. In no European country is there more than one association for a single breed of stock. Although a single association is much more useful, there is in America nothing to prevent any number of breeders who are dissatisfied with the management of their associations from organizing a new one.

Advantage of Single Register.

Practically all associations are chartered in some state and any supervision of their work by state officials is wholly nominal, and the officers elected by the membership are responsible only to the members. Serious abuses have d e v e l o p e d through the acquirement of control of an association by a small number of persons who may direct affairs for their own interest and in a few instances dishonestly. Such mismanagement has been facilitated by the failure of a considerable part of the membership to attend the annual meetings at which officers are required to report upon their administration of the affairs of the society. In other instances interested parties have secured proxies, without instructions as to their use, in sufficient number

Conduct of Herd-Books.

A CLYDESDALE SIX-HORSE TEAM.

to over-ride the number of members usually present in person. The use of proxies is rapidly being discontinued and the more general attendance of members insures fair administration of their affairs. Fraudulent registrations when detected are usually punished by the expulsion of the offender and cancellation of pedigrees of animals owned by him at the time. Some associations are more active than others in detecting frauds, although it is impossible for any set of officers to fully verify the breeding given in each application for registry. The association will be no better and no worse than the average integrity of its individual members.

Imported pure-bred animals have always been admitted to the United States without payment of duty. The law relating to such exemption from duty is a part of the tariff act and has stood for some years as follows:

"Any animal imported by a citizen of the United States specially for breeding purposes shall be admitted free, whether intended to be so used by the importer himself or for sale for such purpose: *Provided*, that no such animal shall be admitted free unless pure-bred, of a recognized breed, and duly registered in the books of record established for that breed. And *provided further*, that certificate of such record and of the pedigree of such animal shall be produced and submitted to the customs officer, duly authenticated by the proper custodian of such book of record, together with the affidavit of the owner, agent or importer, that such animal is the identical animal described in said certificate of record and pedigree: And *provided further*, that the Secretary of Agriculture shall determine and certify to the Secretary of the Treasury what are recognized breeds and pure-bred animals under the provisions of this paragraph."

Until 1904 the Secretary of Agriculture furnished the Secretary of the Treasury "lists of foreign and American books," recording animals entitled to admission to the United States free of duty. In that year an order was promulgated to give some basis of supervision over American associations which had received the certification of the Department of Agriculture. Only such books were placed upon the Government certified list as had complied with regulations established by the Department of Agriculture calculated to insure the fair and proper management of those books. Many associations whose members did not wish to import stock complied with the Government orders and secured certification of their books for the standing it gave them with the breeders and buyers from other countries.

Relation of Government.

In December, 1910, it was ruled that it had not been the intention of the framers of the tariff act to empower the Department of Agriculture to supervise the registration of pure-bred live stock. The certification of books of record was accordingly discontinued. Animals are now exempt from payment of duty only on the certificates of pure breeding issued by the Bureau of Animal Industry of the Department of Agriculture. The officers of the Bureau of Animal Industry define what will be accepted as pure breeding. The customs authorities require that the certificate of registration accompany the certificate issued by the Bureau when the animal comes to port of entry.

The Canadian breeders have worked out a novel and admirable plan of administering the affairs of breeders' associations. The Canadian act providing for the incorporation of live stock record associations was passed in 1900. It provides for the incorporation of such associations upon application to the Dominion Minister of Agriculture. Not more than one association for each distinct breed of horses, cattle, sheep or swine can be incorporated under the act. In 1904 delegates from all existing associations met in convention and organized themselves into the National Live Stock Association. At the same time it was agreed that all existing records should be amalgamated into one National Record. It was also arranged for the Minister of Agriculture to assume the administration of the National Live Stock Record. The various associations retain their identity, continue their work of promoting breed interests, make their own rules, and elect a member of the joint executive committee known as the National Record Board. This board deals with matters in which the societies are jointly interested. The record offices are in the government buildings and each certificate of registration is examined by a representative of the Minister of Agriculture and if approved has affixed to it the seal of the Department of Agriculture.

Canadian Registration Affairs.

Nearly all records in all countries limit registration to the offspring of registered parents. Newer breeds have less rigid standards agreed to until such time as it becomes advisable to receive no more foundation stock. The American Trotting Register and the American Saddle Horse Register are to be closed to all but progeny of registered stock in

Eligibility to Registration.

1913. All such regulations, charges for registration and disposition of accumulated moneys are decided upon in business meetings of the members. Membership fees vary from $1 to $20 per year. Any breeder may become a member of the association of the breed he handles and as a member has a vote on all questions and is eligible to hold office in the association. In many associations the charges for registration are lower on animals owned by members than on those owned by non-members.

Breeders organize for other purposes also than mere registration. In many instances standards of excellence have been prepared and distributed *Other Functions of* to secure uniformity in the objects *Breed Associations.* of the breeders. Expense is sometimes incurred in advertising and in arranging for sales of stock in districts that give promise of developing a demand for breeding stock. Of recent years the associations have appropriated large sums to be offered as additional prizes at important fairs. The extra premium money insures larger and better exhibits to make a creditable display to attract the public and to acquaint them with the accomplishments of the breeders and with the value of superior stock.

The discussion so far has dealt primarily with the fundamental principles underlying successful breeding. The aim has been to present the known factors which determine heredity and to give an unbiased summary of the theories advanced by noted scientists for explaining the mysterious features of reproduction. The molding of animal form and function is an art in itself, based mainly on the practical experience and achievements of the master breeders but aided in later years by the discoveries and theories of scientific investigators. In the light of the preceding pages, a brief consideration of the steps by which horses, cattle, sheep and swine have reached their present state of perfection and adaptation to human needs will give a clearer understanding of breeders' problems.

CHAPTER XIX.

HORSE BREEDING.

The faithful horse has been the subject of a great many unfulfilled prophecies. At different times his friends have thought the time of his passing had come but the factors that seemed to threaten his existence really enlarged his field. Up to the latter part of the last century every new application of mechanical power seemed to promote trade. The number of wealthy persons grew and one of the first manifestations of that wealth was the family carriage, appropriately horsed and stylishly equipped. The use of the automobile as a pleasure vehicle and as an evidence of wealth has vitally affected the outlet for light horses. The draft horse, on the contrary, was never before in such great demand.

In the first part of the nineteenth century cities were comparatively small and business not highly organized.

Place of the Draft Horse. The chief demand for large work horses came from the managers of wagon freight lines. Shipping of merchandise from eastern cities by canals in 1825 and by railroads in 1835 appeared to have done away with the greatest need of the horse. In reality, these changes brought with them the greatest commercial demand for horses, improved means of travel and transportation, extended commerce and stimulated the growth of cities. The shipping and distribution of manufactured articles has always made the draft horse the peculiar need of the large business concerns in our modern cities. It is the city demand that has given the draft horse his place in America. The same is true of other countries.

The future will doubtless bring a more general appreciation of the economy of the larger horse in farm work. His use on the land means larger implements and less man labor per acre. Farmers of hilly lands may continue to use lighter horses or to adopt the larger ones and lay greater emphasis on freedom of action. It is still true however that the price of horses purchased for farm work is on a level with the earning capacity of those animals on the city streets. The city is the chief consumer and the values it places upon the various types and sizes are the ones that govern at all points where forces are dealt in.

Cities Set Prices.

By 1850 the development of American cities and the organization of business was such as to produce a demand for work horses of greater size than had previously been in use. Until then the horse stock of the country was chiefly related to lighter breeds. A few English and Scotch stallions and some with French blood may have been in service in our central states. These would have come chiefly through Canada to satisfy such demand as existed for farm horses of greater weight. In 1851 the first Percheron came to Ohio. His colts were so highly esteemed that he was soon purchased by Illinois breeders who were anxious to

Earlier Draft Horse Affairs.

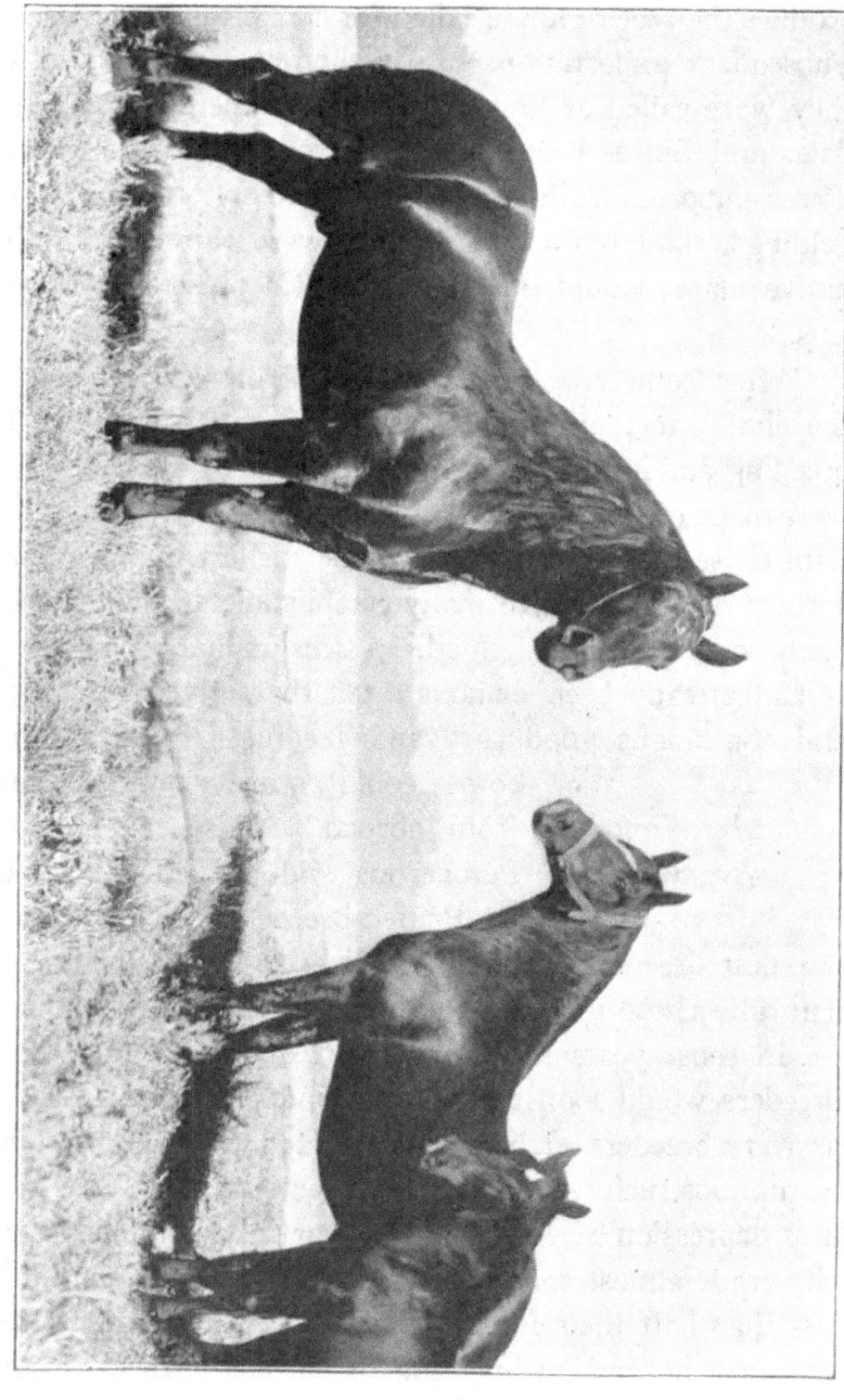

A PERCHERON FAMILY.

produce the type of horse called for by Chicago team users. Subsequent importations of Percherons, or Normans as they were called at first, were quite numerous. Clydesdales and Shires began to come from the British Isles. These imported stallions were used chiefly to sire market geldings, though the largest that were raised from the native mares would only be chunks in present-day markets.

After commerce had recovered from the effects of the civil war there was a period of greater activity in breeding and in the eighties importations of all the breeds were made on a large scale. Foreign-bred mares, together with those that descended from the earliest importations, formed a foundation in many establishments that seemed likely soon to furnish home-bred sires of market geldings It had already been demonstrated that, given the blood and continuous good care and feeding, American-bred horses could compete with the best

American Breeding. from abroad. This was true of Percherons and particularly so of the British breeds, because many of our best breeders were of English or Scotch birth and naturally chose to breed the horse of their native soil.

In those years it seemed that American draft horse breeders would soon be as nearly independent of Europe as were breeders of beef cattle. Had there been no interruptions such would have been the outcome. The business depression of 1893 and the year following closed the city trade almost completely. The best horses shrank to less than half their former values and the fortunate man was he who had no breeding or surplus horses. Regis-

PERCHERON STALLION.

tered mares were sold for what could be had and, in the frantic efforts to realize something from what seemed a wreck, nearly every horse-breeding enterprise was abandoned and the results of slow and expensive efforts, by which our breeding was about to be established, were wantonly sacrificed. The occurrences of that period are responsible for the fact that today, sixty years after the first importations of draft stallions, the bulk of the draft horses marketed are the progeny of sires purchased in European countries.

The Depression.

The wonderful expansion of trade that began in the closing years of the nineteenth century found our farmers unprepared to supply the great numbers of high-class horses needed for city work. There has been a scarcity of mares of sufficient size and breeding to produce real drafters even when mated to the excellent animals included in the large numbers of stallions brought from abroad in recent years. The scarcity of mares likely to raise stallions fit for stud service has been still more marked. The imported stallion is so common and the home-bred so often inferior by comparison that it has too often been granted that superior horses cannot be raised in America. The fact is that the main if not the only real advantage enjoyed by breeders abroad is in the superiority of their mares. They have been breeding along the same line ever since the draft type was established. They have not needed to put size before all else

The Revival.

Advantages of Foreign Breeders.

CLYDESDALE STALLION.

in choosing stallions to breed to. Their mares have been as large as desired and the selection of stallions has been based upon conformation, soundness, action, limbs and feet and other features for which our breeders are so ready to pay large prices. While a large proportion of the winning mares and stallions in our draft-horse shows are imported, American-bred winners have been numerous enough to demonstrate that the right breeding and right feeding are as successful here as anywhere. Our home-bred winners are the offspring of mares whose ancestors have been selected for several generations for quality and action no less than for size.

Home-bred exhibits have scored more heavily in the mare than in the stallion classes. Of late years practically all mares eligible to registration have been retained on the farms. Because young stallions are so much more troublesome only a few are kept entire and those that are spared are much less likely to receive the feed and care essential to the development of draft qualities. It does not seem likely that horse breeding will be taken up as a special business as is the breeding of other classes of stock. The risk of mares failing to raise foals makes it desirable to obtain some work from them to pay their keep when not breeding. A considerable amount of work can be done without injury to mares even when they are breeding. Such work must be done in the hands of a careful teamster and many failures to raise foals are attributable to the driver of the mare. For these reasons draft horses are likely to be raised by farmers rather than

Stallion Raising.

SHIRE STALLION.

by professional breeders. The methods coming into use for getting mares in foal by artificial means are sure to add greatly to the returns from horse-breeding and to make both mares and stallions more profitable as investments. Farm breeding is much more favorable to raising mares than stallions though this is not necessarily true. It is possible that America may develop the French custom of buying numbers of colts at weaning time to be developed on separate farms until salable as breeders.

The continued high prices for horses have helped to direct the consideration of team users to the auto-truck.

Influence of the Auto-Truck. That a part of the work now done by horses can be done by mechanical power has been demonstrated. The relation of costs of hauling by horses and by motor power is not so clear. Considering the amount of work that must always be performed by horses and the continual expansion of commerce there is every reason for assurance that really high-class horses will command more than it costs to produce them. The competition of the auto-truck or any other factor that tends to change values affects first and most seriously the animals of lowest earning capacity.

Breed for Top of Market. The smaller sorts or those that lack durability because of defects in foot, limb or body will be the first to be discriminated against. To insure maximum returns it is therefore always a practical economy to breed for the top of the market, to select parent stock and care for them and their offspring in such a way as to reduce to a minimum the probabilities of having misfits to sell.

BELGIAN STALLION.

Buyers of draft horses are as far from being unanimous in their ideas of perfection in form as are breeders. Points of conformation offer little ground for divergence of opinion, but there are at least two types either one of which will suit the preferences of different buyers. It is of small moment whether one breeds for the upstanding free-going kind or for the lower and extremely drafty type that matures earlier and is of more phlegmatic disposition. The former type is commoner in the Clydesdales and Shires, the latter among Percherons and Belgians. The British breeders argue that while their horses are somewhat slower in attaining their maximum development, their fibre is such that they remain serviceable through a great number of years. The continental breeds find favor with farmers by coming onto the market while still quite young. All of these peculiar features are as much matters of individuality as of breed.

Draft Types.

Since the revival of the draft horse demand size has been at so great a premium that there has been but little discrimination among different kinds of horses that are up to required size and weight. Firms buying geldings for show and advertising purposes have bought quality and action at a high premium, but in the main softness of bone, coarseness of joints and defect of action have been very lightly discounted. It is impossible that animals with such characteristics can do as much work or remain in service so long as the better sort, but in the activity of the demand for

Market Discrimination.

HACKNEY STALLION.

weight there has been little evidence of observations of difference in the wearing of different kinds of horses. It is inevitable that discrimination will soon be made in the selling ring as judges have been doing in the showring. Size will be a prime requisite no less than it has been, but cleanness and correct set of limbs and trueness of action directly related to a horse's value on the street will be appreciated whenever the buyers are sufficiently independent to be discriminating. In service as in breeding individuality is more of a controlling factor than is the breed name.

Breeding the lighter classes of horses is not so strictly an agricultural matter as is the breeding of drafters. Good driving and saddle horses continue in demand and at prices that make their raising an attractive business. Breeding is even more of a factor in light-horse breeding than with draft horses, at least feeding is less to be relied upon for producing market qualities. Neither coach, driving nor saddle horses enter into commerce to any considerable extent. Their users are persons who own horses for the pleasure of using or showing them, consequently there is practically no limit to the amount obtainable for finished specimens, and misfits or sub-standard sorts are unlikely to bring more than the cost of rearing them. Some of the lighter less level soils in the localities of larger centers of population are better adapted to the rearing of light horses than are most of the farms in the cornbelt.

A carriage horse seldom develops sufficiently before five years of age to show the style and action that sell best. Even the most promising young horses require a

STANDARD-BRED STALLION.

year or more of expert handling and most of the more successful show horses are handled through a number of seasons. Such handling is expensive but when a show horse results the expense is fully justified.

Breeding Carriage Horses.

In the past many of our best specimens of high-stepping horses have been found by dealers who handle large numbers of prospects in the hope of developing a few of extraordinary merit. The business is fast becoming more fully systematized. If extreme style and action are not transmitted with great certainty it is largely because matings are rare in which both parents are backed up by ancestors that were themselves of show caliber. Raising coach horses is not an attractive business to one who is not prepared to develop and market the horses he raises. The breeder of a horse has the best opportunity to develop him fully and if he does not do so he can receive only fair prices for those he sells as show prospects. The trouble and expense of developing is justified only when the breeding is of the very best. It is not sufficient that there be uniformity of conformation in the breeding stock, but there must also be similarity in the spirit and ease and height of action that go to make up the highest type of carriage horse.

Good Breeding an Essential.

Breeding Trotters.

The outlet for roadster horses is not so broad as before the coming of the automobile. Really high-class road horses are wanted however, and the demand for horses likely to develop into racers is unlimited. The sport of trotting racing is firmly established in every part of our country. Horse racing is in no sense an agricultural affair, but selling young animals for driving purposes or to be prepared for racing by those engaged in that work offers an opportunity for profitably combining good farming with ability as a breeder. Here also, while individuality counts for much, performance is the main thing and it must come by inheritance. To secure profitable prices for undeveloped trotters they must be correctly built, faultless in action and above all bred in lines that have produced extreme speed.

CHAPTER XX.

CATTLE BREEDING.

Improved cattle were brought from Europe to America in the latter part of the eighteenth century. There have been periods of depression when scant progress was made, yet on the whole advancement has been steady and any one of the leading breeds could continue to go forward without further importation. The Short-horn was strongly entrenched in most parts of the country before its rivals appeared. American-bred Bates Short-horns were purchased by English breeders in the seventies. The opening of the range trade in the late seventies diverted attention from the deep-milking beef cow then so widely distributed on central and eastern farms. The ranges furnished feeding cattle more cheaply than they could be produced under the best systems of farming. The only opportunity for the breeder was to furnish bulls for the western steer-raiser. That trade discriminated against heavy-milking qualities. It also prized the ruggedness and heavy-fleshing qualities of the type of Short-horn then being brought from Scotland and the Bates era closed with the beginning of the Scotch boom.

Influence of the Range.

The western trend of population has now removed some of the economic advantage enjoyed by the ranch-

SHORT-HORN BULL.

man in the cheapness of his land, and a return to farm breeding is certain. The early maturity and heavy fleshing of the Scotch type are even more valuable than formerly, but strict economy in producing beef on high-priced lands is reviving the demand for milking qualities in the dams of steers that are to be fed out by the men who breed them. British farm conditions are very similar to those that exist in much of the territory where Short-horns are most numerous in America. The British standard has included the qualities needed in America and our breeders have continued to resort to the older herds over the water for herd bulls and for occasional females though the need for these is declining as time goes on.

Beef and Valuable Lands.

The popularity of the Hereford with the range trade in the eighties had much to do with stimulating the change of type of Short-horns. The Hereford had always been bred for grazing purposes and was little affected by the transfer to the grass lands of the West. Importations were numerous in the eighties, but in the nineties it became apparent that our own breeders were more successful in improving the weaknesses of the Hereford than were the friends of the breed in England. Since then there has been scant need for importing and it is admitted that this breed has been improved in this country more than any other imported breed of cattle. Hereford interests have not been hampered by any color craze, and family names have always been less considered than tangible merit in nearby ancestors. The

American Progress in Herefords.

HEREFORD BULL.

early-maturing feature has been fixed in them also and their easy-fattening qualities have added to their popularity among feeders of young cattle, though the same feature has worked against them to some extent when the feeding periods have been of longer duration.

The Angus has less numerous advocates among the ranch men than have the two breeds referred to but the prices Angus steers command from cornbelt feeders win them very ardent advocates. The coming of this breed followed closely after that of the Hereford and it is strongest in sections where cattle feeding is carried on extensively. The peculiar advantage of this breed is shown in its wonderful record of winnings in block tests and in competitions for prizes for finished bullocks. The Galloway enjoys almost undisputed sway under climatic conditions resembling those of its native home in southwestern Scotland. The enterprise of its breeders is bringing this breed the appreciation that is its due.

A few factors have operated upon all breeds alike during the last two decades. The changing tastes and style of living of our population make themselves felt upon the markets through the purveyors of meats. As a result the live-cattle weight of greatest popularity has steadily declined. The other factor that has operated in the same direction is the demand for early maturity. The greater economy of production in young animals has been more and more appreciated as meat-producing lands advance in value and require the most economical utilization of the crops produced. The result has been that the modern beef type is practically the

Evolution of Types.

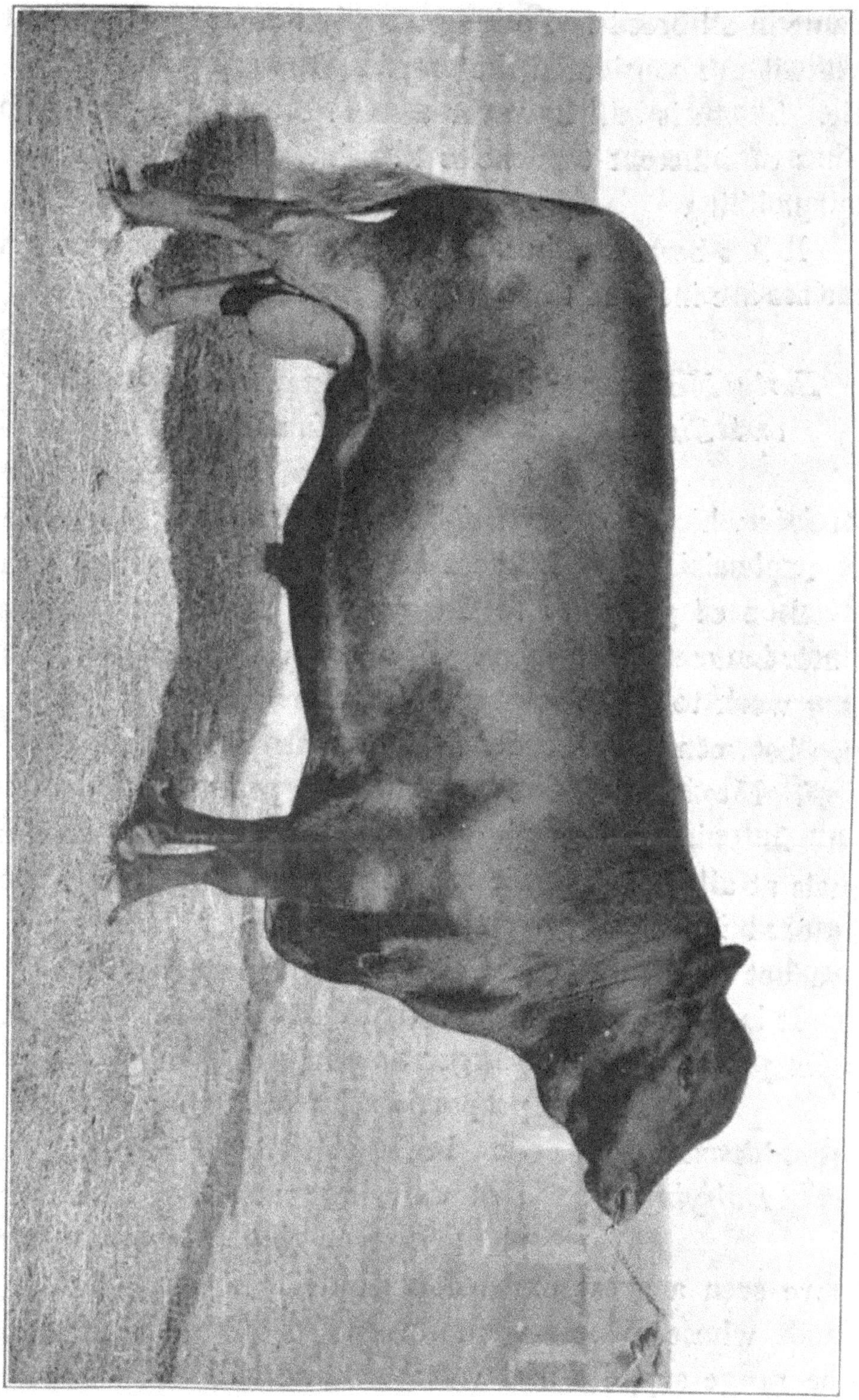

ABERDEEN-ANGUS BULL.

same in all breeds. So long as only the really good individuals are considered the breed means comparatively little. There is still however a wide divergence along the lines of inherent tendencies that have much to do with adaptability.

It has been true in all stock breeding that breeders as a class are inclined to go to extremes. Since the majority of all the breeds have come to be of the smaller more early-maturing stamp, a question arises as to the effect of continued breeding from stock in which size and weight have often been overlooked in emphasizing smoothness of build and fleshing. The relations of gains to feed consumed is the fundamental factor in meat production. Breeding for the market alone may work to the detriment of the producer and create a gap between the breeder and his patrons, the raisers of beef. Maximum gains consist of the products of growth and fattening. Growth is less pronounced in small animals at all ages and the relation to size cannot safely be ignored in efforts to combine the highest value of the product with the greatest economy of production.

Early Maturity and Size:

It is commonly admitted that cows bred especially for milk production yield larger amounts of human food in proportion for feed consumed than do beef cattle. This suggests that dairying must inevitably supplant beef-making. Recent years have seen a great expansion of dairy interests. Many lands whose owners were forced out of beef-raising by the range supplies and were unsuccessful in speculative

Advance of Dairying.

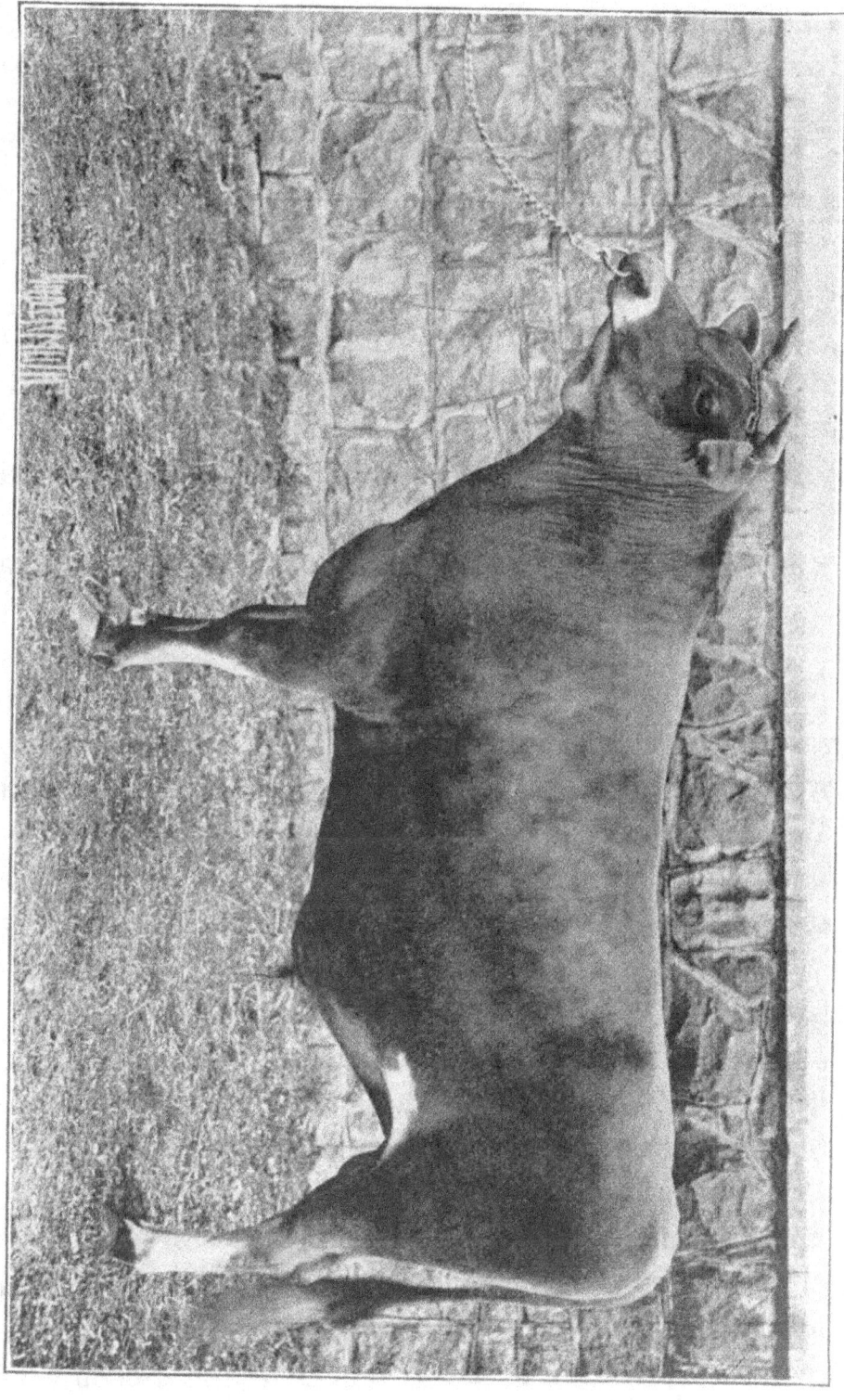
YOUNG JERSEY BULL

feeding operations have become badly run down. There has been a great increase of consumption of milk and butter proportional to the growth of population. Supplies could come only from the lands located near the markets for dairy products. Dairy animals are now established on many lands that had carried little stock since beef-making was relinquished.

It is important to recognize the fact that the relation of product to feed is more easily studied in feeding for milk than when meat is a marketable product. Partly for this reason and partly because beef-raising has been regarded as the business of cheap lands, the feeding of beef animals has not been made so scientific as has the work of the dairyman. The individualities of animals have not been so highly regarded nor so closely studied by those engaged in meat production. There is probably less variability in productive capacity among animals bred in beef lines than exists in milking stock. The abandonment of wholesale methods will decrease the difference that has seemed to exist between the possible returns from these two classes of cattle.

Advantages of Dairying.

The general adoption of dairy farming in older parts of the world is evidence that this kind of cattle husbandry must ultimately prevail. In such countries the labor question has a vastly different aspect from that which it presents in America. The eating of meat will continue so long as it can be purchased at prices within reach of the bulk of the population. Its production is less dependent upon the labor factor than is that of milk. Scientifically

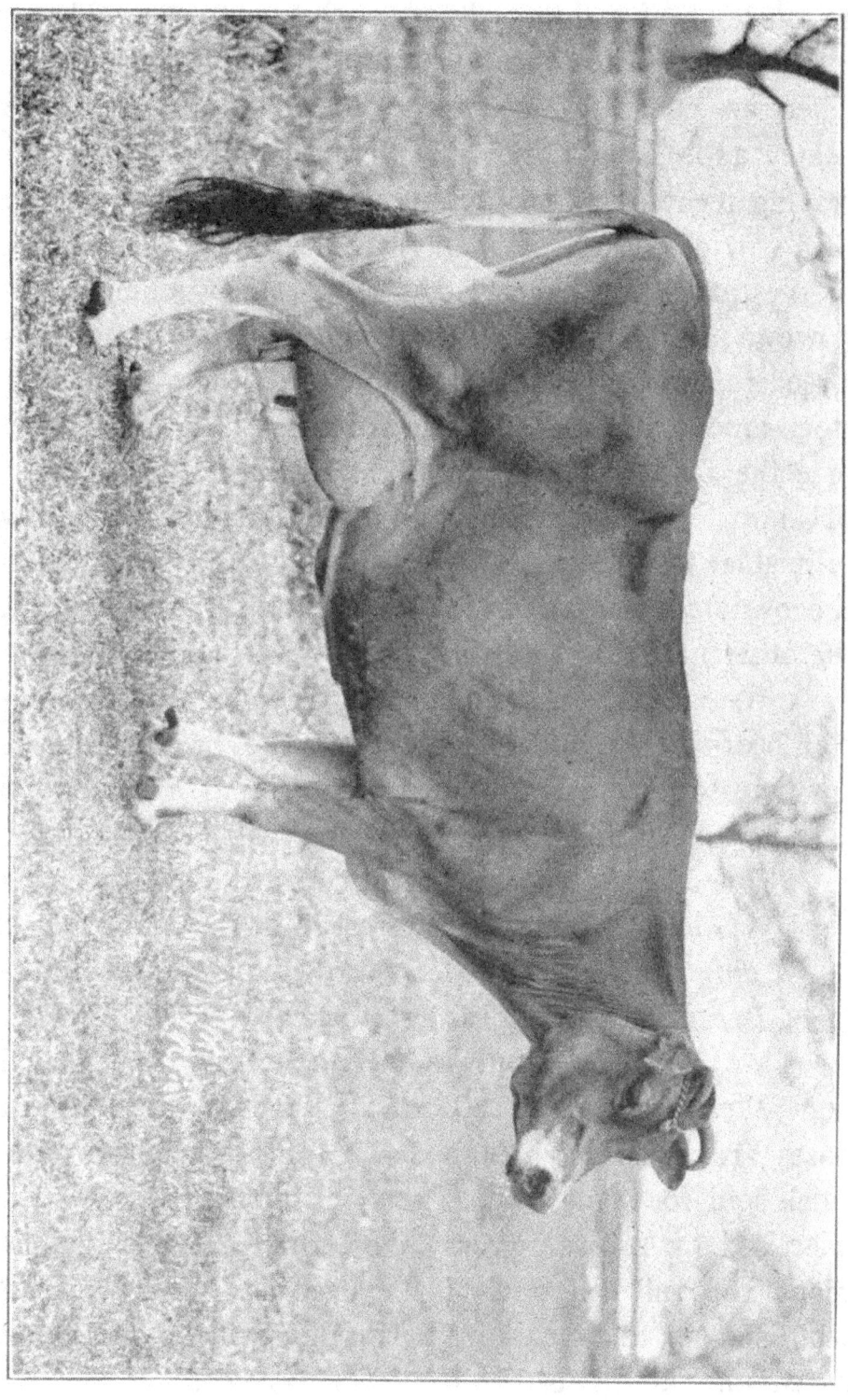

JERSEY COW.

conducted beef-making will offer rewards to careful breeding and studied feeding for an indefinite number of years. There are considerable stretches of country in the United States and Canada that can be utilized successfully in rearing feeders and stockers under a system that requires of a cow only that she shall rear a calf each year. Most of the cattle must be raised in the grain-producing areas however, and all of the fattening must be done there. Breeders who expect to find an outlet for their surplus stock among farmers of these valuable lands must recognize the fact that the need is for a profitable beef type of animal with sufficient milking capacity in the cows to enable them to raise two calves each year or to allow the owners to sell the milk of one-half the herd while the other half raises all the calves.

In Great Britain experience has shown that commercial beef-raising and the work of the breeder do not combine satisfactorily. The fattening and marketing of less valuable animals is no hindrance to a breeder's work, but when the main interest is in the market stock the details of rearing and selling breeding stock are not likely to be fully looked after. Unless that is done, the returns from sales of animals for breeding purposes are likely to include less profit than those from stock sold for slaughter. The discrimination likely to be exercised by future buyers of breeding animals will demand the full exercise of the breeders' skill while he who chooses to breed for the market will be the gainer by paying fair prices for the bulls he needs.

Professional and Commercial Breeders.

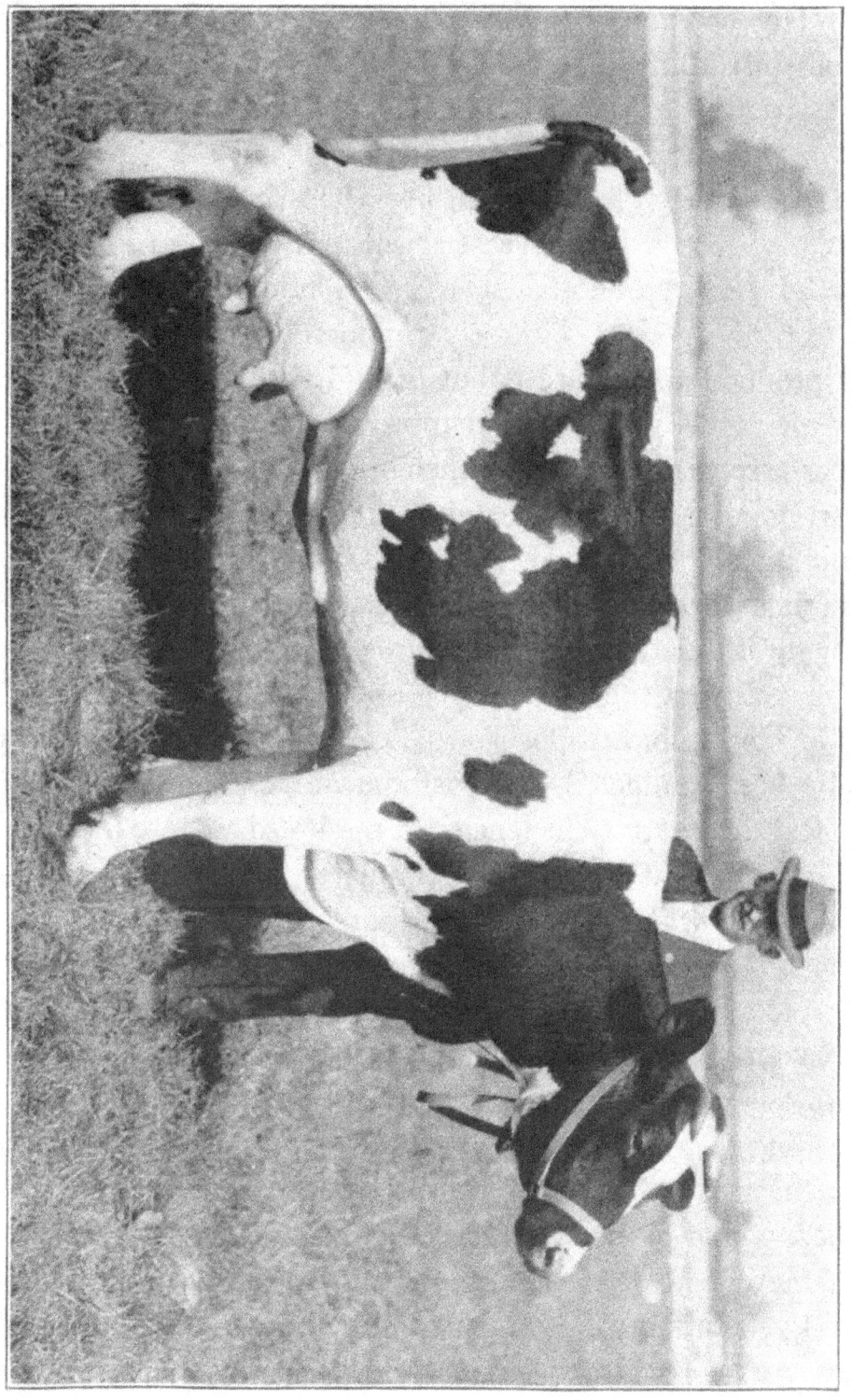

HOLSTEIN-FRIESIAN COW.

The breeding of dairy cattle is comparatively free from disturbing elements. The principal variation is in the strength of demand for the more meritorious stock.

Superiority for Dairy Purposes. The superiority of the improved animal lies almost altogether in her greater economy of production. Cost of production is usually inversely proportional to the amount of production. There is but little difference in the value of the solids produced by improved and by scrub cows. The breeds differ among themselves mainly in the proportion of solids to the whole amount of milk yielded and in adaptability of the solids to various uses. The animals themselves differ in adaptability to conditions, chiefly in relation to the amounts of roughages and concentrates that they can most profitably utilize.

Public competitions and so-called breed tests such as have been conducted at expositions are apparently planned to test the relative efficiencies of the breeds entered. Such

Breed Tests. trials stimulate interest and encourage study of breeding, feeding and management. As competitions they serve to test the skill of the various feeders. It is doubtful if any other branch of the feeder's work is carried on so successfully in America as is the feeding for milk production. Few of the records made in the native homes of the breeds are comparable with those found in the Advanced Register and in the Register of Merit. Most of the high producers trace through several generations of American-bred stock. Jersey breeders import more largely than do the workers with other

GUERNSEY COW.

breeds. The Island cattle are highly bred in producing lines, but much of their popularity is due to their attractiveness of form and symmetry which have often been ignored by American breeders in basing their selections solely on features indicating producing capacity.

There is no need for speculation as to the future place of the dairy cow and but little occasion for discussion regarding the type of the future. The most useful and the most highly prized animals will always be those that produce most largely and most economically. There seems but little to be desired in realization of greater possibilities though doubtless records still more astonishing will be made. Adoption of the plan of reporting records of the feed consumed at the same time the milk and butter yield is reported will facilitate closer study of economy of production. The accomplishment most to be hoped for is the fixing of the qualities of the record-makers so as to raise the average records of the breeds. In selections for this purpose the cows' records as producers and breeders and the bulls' records as sires of tested daughters are of extreme value. Yearly records are much more useful in this connection than those covering shorter periods. Some breeders, among them many successful ones, assert that in breeding for production the records may safely be made the sole guide in selection. In such a study we feel the weight of the handicap of ignorance regarding external evidences of digestive power and milk secretion.

Advances in Dairy Breeding.

Form and Function.

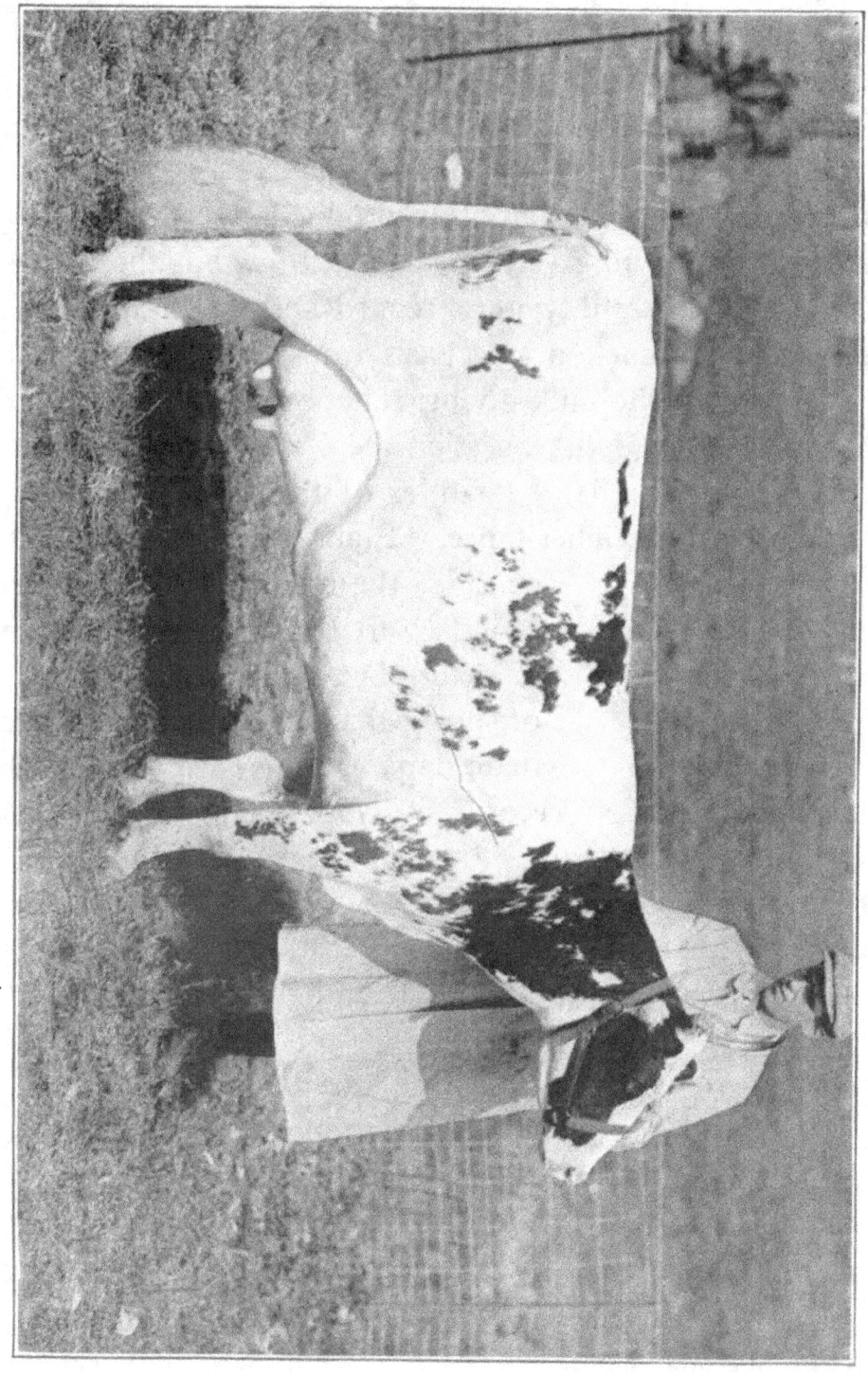

AYRSHIRE COW.

The relation of form to function is very imperfectly understood though it is known that certain physical qualities are uniformly characteristic of proved cows. Breeding only by the aid of records may often suggest combinations of blood lines in each of which there is carried a fault that limits capacity for production. Breeding by records alone will give results for a time, but the highest achievement in utility must regard both form and measurements of function as a basis for selection.

Although the milk-giving function is undeveloped in the males there is no question as to a bull's capacity for transmitting the dairy qualities of the females that contributed to his inheritance. Some successful breeders consider the record of the paternal grandam to be of greater moment in determining the capacity of the offspring than is that of the immediate dam. If this idea can be substantiated by test records it would suggest an extra strength of inheritance through the males of the dairy breeds, not necessarily because they are males but because more carefully selected and more strongly bred than is usually the case with the females of the herd.

Extra Influence of Sire.

It is impossible to disregard the possibility of impairment of vigor in the offspring carried by cows that are at the same time being stimulated to their highest milking capacity. If the foetus is not more than three or four months' old when the testing of the dam is discontinued there need be no serious result. If it is older and the dam responds

Testing Breeding Cows.

to feeding by an unusual yield of milk it is more than probable that the offspring will be weaker than it would have been if the needs for foetal development had been given primary consideration in adjusting the ration.

CHAPTER XXI.

SHEEP BREEDING.

Sheep breeding presents difficulties and offers rewards not met with in other classes of stock. One of the very considerable rewards is the pleasure of having overcome the difficulties. The peculiar need of controlling conditions in order to make sheep-raising a success grows out of the fact that the environment of farm sheep is highly artificial. The undomesticated sheep is a mountain dweller. In its habitat it is free from all dampness and always able to graze over large areas to secure the variety of vegetation necessary to appease its appetite. This peculiar appetite is the thing most generally prized, because it makes sheep valuable as weed destroyers. When confined to farm pastures they are soon deprived of the variety of food for which they crave and compelled to pass many times over the same ground. Under such conditions, and particularly when the soil is not of a dry character, there is a fostering of pests and diseases that are very difficult to treat. These troubles of farm sheep that are so difficult to cure are preventable if the essentials of the environment are made more nearly comparable with those of natural conditions. On smaller farms this can be done by grazing the flock upon a succession of green

Environment for Sheep.

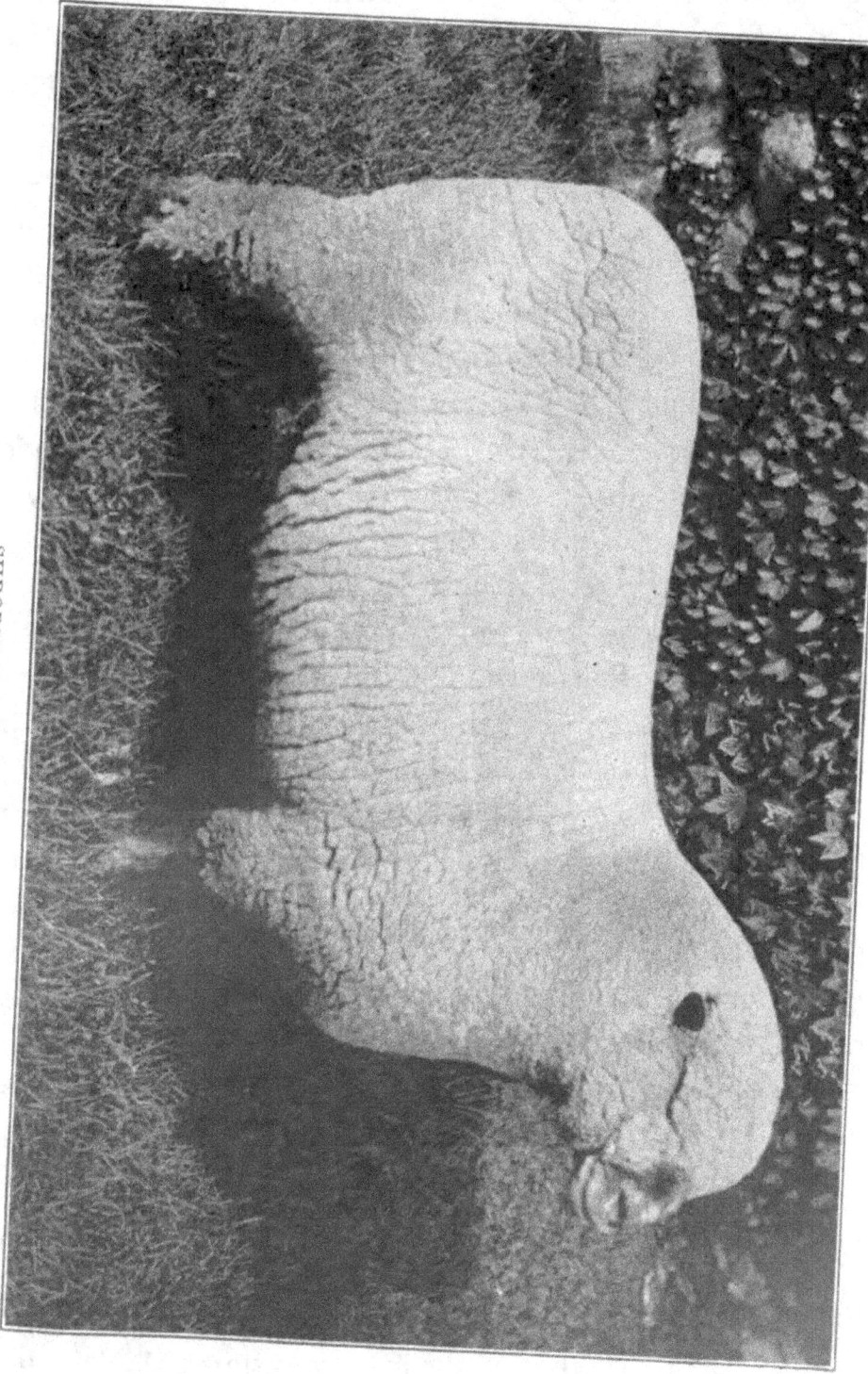

SHROPSHIRE RAM.

crops that can be arranged to furnish a variety of forage and sufficient change of ground. Such farming for sheep involves more labor than allowing them to run on old pastures but it wards off their peculiar ailments and permits them to make their maximum returns. Furnished with the right kind of feed sheep will consume more in proportion to their size than do large animals and produce more in proportion to the feed consumed. It is only when the essential features of natural environment are preserved and improved under domestication that sheep can thrive fully. Otherwise they are less useful than they may be.

Economy of Sheep Raising.

All of our breeds of mutton sheep originated in England. Much has been done in fixing characters in the various breeds that will adapt them to varying altitudes and types of soil. The length of time through which the oldest of the breeds has been selected and cared for is far too short to overcome the force of natural features bred into the original stock by thousands of years of natural selection. Mutton production is profitable under the careful English shepherding on valuable lands. Imported sheep are much more prominent than home-bred ones in American shows of the mutton breeds and large numbers of rams are imported to head breeding flocks.

English Shepherding.

The unusual development and the vigor of the English sheep is a result of the system of rearing their stock. Climatic advantages enable their shepherds to raise the crops needed with fewer difficulties than are met with in some states, but it has

SOUTHDOWN RAM.

been abundantly proved in our shows that good breeding supported by the right kind of care and feeding produces sheep in this country fully equal to the best from the homes of the breeds.

Buyers of breeding sheep are seldom prepared to estimate the merits of pedigrees. Selections are based nearly altogether on individuality alone and this explains part of the disappointments that are not uncommon.

Pedigrees Not to Be Ignored. When selections are made in such a way as to secure actual merit supported by breeding, and the details of care and rearing are attended to, sheep breeding is a most pleasant and profitable occupation. The value of good blood is often obscured by the fact that the lambs from meritorious sires and dams are not handled in such a way as to permit them to exhibit the capacity for development that they have inherited. Success with sheep depends upon unremitting attention to a number of details, and the more intensified farming of the future is certain to bring in a more general and more careful sheep husbandry.

The numerous breeds of medium and long-wool sheep represent every combination of qualities likely to be needed for adaptation to particular sections or systems of rearing. They differ in size and in character of fleece, but the more vital distinctions grow out of the factors that have governed the selections of the makers of the breeds. Selection for rapid growth in some breeds has necessitated less strict adherence to the mutton form than has been practiced in the homes of other breeds. The fundamental features of adaptability are revealed by

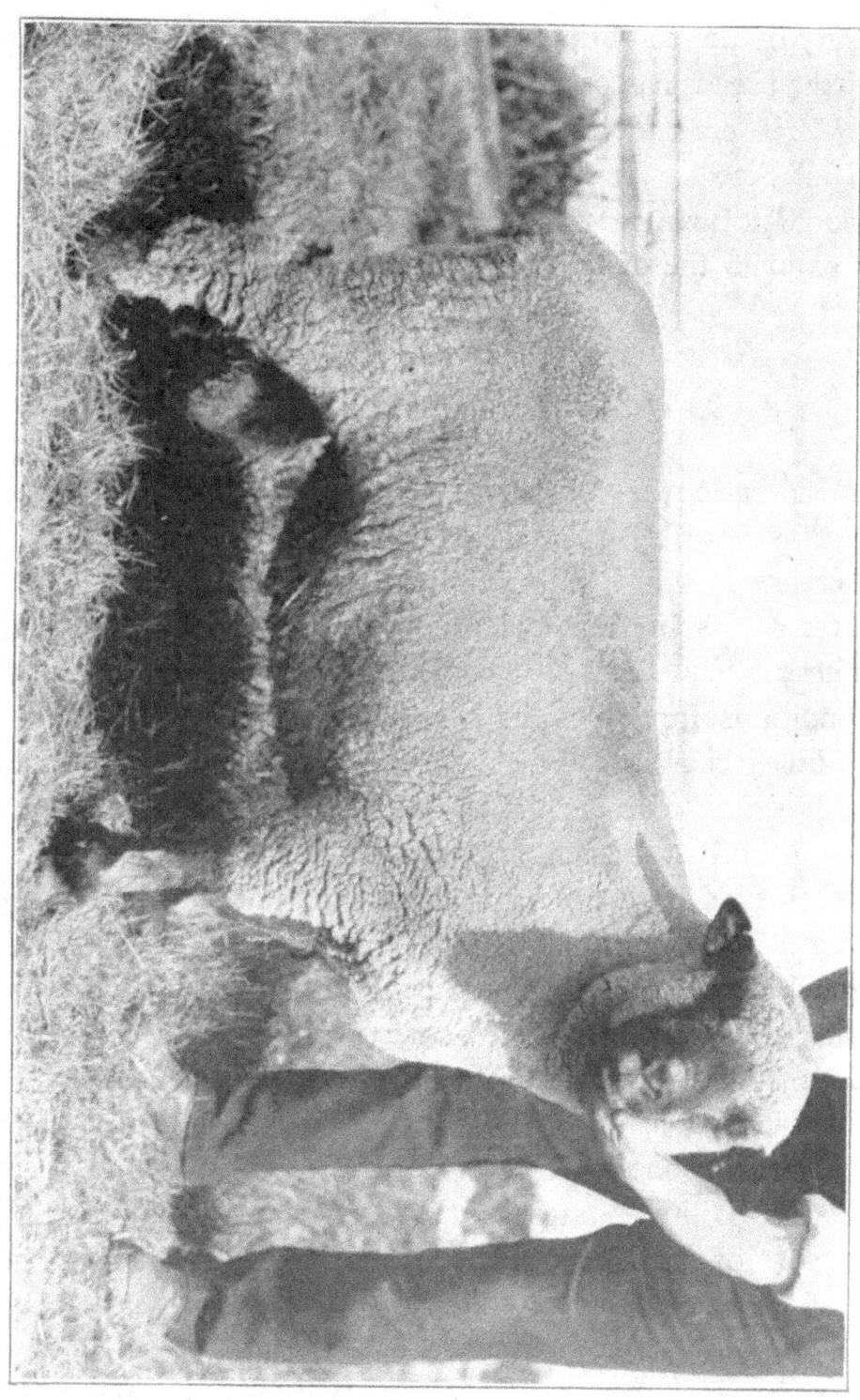

HAMPSHIRE RAM.

study of the conditions under which and for which each of the breeds has been developed.

Breeders and judges magnify the importance of type in all breeds. It is sometimes insisted that an animal should not win, no matter what its mutton qualities, unless it exhibits the distinguishing characteristics of its breed.

Breed Type. These characteristics are considered to consist mainly of the covering and features of the head, and general conformation is given secondary consideration as contributing to type. Although breed type is very desirable it is not an end in itself. The measure in which any animal exhibits the peculiar features of its breed should indicate its possession of the inherent tendencies that constitute its adaptability to the conditions for which its breed was produced. So long as breed character is held secondary to mutton qualities in breeding a mutton flock these incidental peculiarities are useful as indicating trueness to breed usefulness. When type is construed to consist of minor peculiarities of head and coloring, and more vital qualities are relegated to second place, then the indication is substituted for the reality, and actual commercial usefulness must decline. Breed points, or fancy points as they are sometimes erroneously designated, have a value but it is always a secondary one.

The United States is still a large importer of wool. A part of the home-grown wools are the equal of the best produced elsewhere but the amount has never equaled the requirements. Cheaper lands in countries of sparse populations produce the wool needed by older countries.

COTSWOLD EWE.

Our own southwestern and northwestern states rank high in wool production but few of the farming sections have entered extensively into the production of wool. In 1807 the states then formed offered bounties to encourage the production of wool. Societies were also formed at that time to encourage all kinds of home manufacturers and render the nation less dependent on materials from abroad. In the same year the first Merino sheep were taken west of the Alleghany Mountains. These sheep were the offspring of stock reared in Spain. Except for short periods wool-growing has ever since been fostered by tariffs designed to keep wool prices above the values in the countries where they are produced more cheaply. When such protection has been temporarily withdrawn, the breeding of fine-wool sheep has been seriously affected.

Fine Wools in America.

Fine wools are used in making of fabrics that could not be made from the wool of the mutton breeds. When fabrics made from Merino wools are in light demand the price of the wool is not in proportion to its quality but the weight of the fleece is always an important factor. Numerous sub-breeds and classes have been produced and provisions furnished for registration of local types and strains not sufficiently distinct to be considered breeds. There is great difficulty in maintaining the maximum density and fineness of staple. For this purpose some flocks have been maintained to furnish these qualities in an extreme degree even though the sheep themselves are not considered suitable for farm breeding. Although it cannot be denied that these features are very difficult to

AMERICAN MERINO RAM.

276 BREEDING FARM ANIMALS

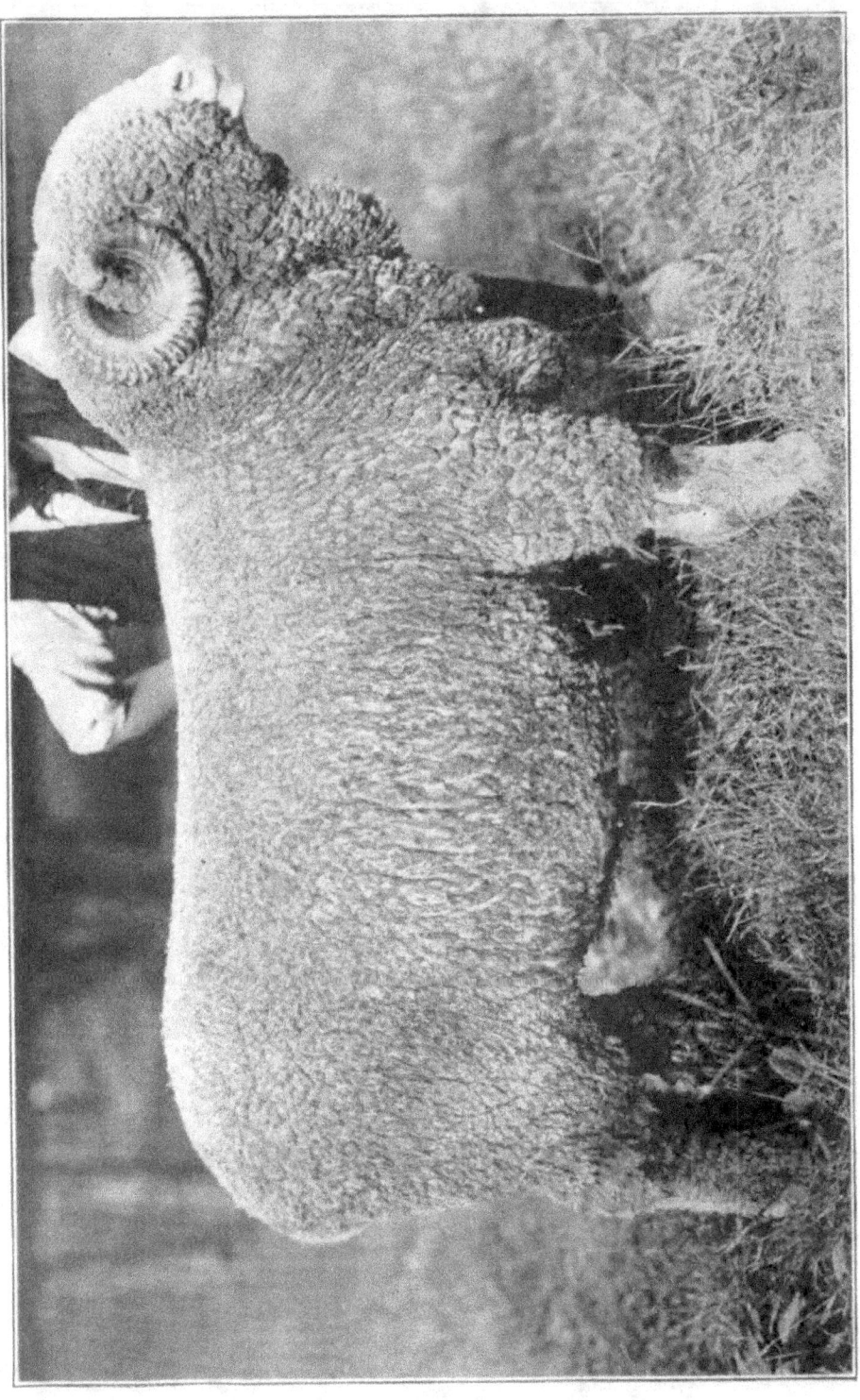

RAMBOUILLET RAM

retain, it is true that part of the difficulty has been due to lack of appreciation of the influence of ancestry. The offspring of sheep of the desired practical type have shown a deterioration due in many cases to the fact that the parents themselves or the grandparents were the result of the mating of extreme types, and then particular individualities were not fixed enough to insure their transmission.

Maintaining the Type.

Considerable numbers of sheep of the fine-wool classes have been exported to South Africa and Australia in recent years, and even should commercial breeding for wool be interfered with, the prestige of our professional breeders should enable them to continue to breed for the foreign trade.

CHAPTER XXII.

SWINE BREEDING.

The swine-raising industry has reached a development in America greater than anywhere else in the world. Other countries have effected improvement in their bacon-producing swine, but the world's lard supply comes from the cornbelt. The improvement of swine for lard production began with the occupancy of cornbelt lands. The horses and cattle brought by settlers from east of the Alleghanies and from Europe were satisfactory for the time, but it was not so with the swine. Upon this class of stock devolved the work of readily converting the easily grown corn into a marketable product.

Since the establishment of the first American breed, the Poland-China, down to the present, the most serious problems of American swine breeders have arisen from that striking feature of the environment of their swine, the corn diet. The original types, of which there were many, were too coarse and ill-proportioned from the standpoint of those in charge of the first packing enterprises. It was also clear that a greater economy of production was desirable. Development was too slow and too small in proportion to the feed consumed. The offspring of some of the stock at hand became marketable at an earlier age

Need of Improvement.

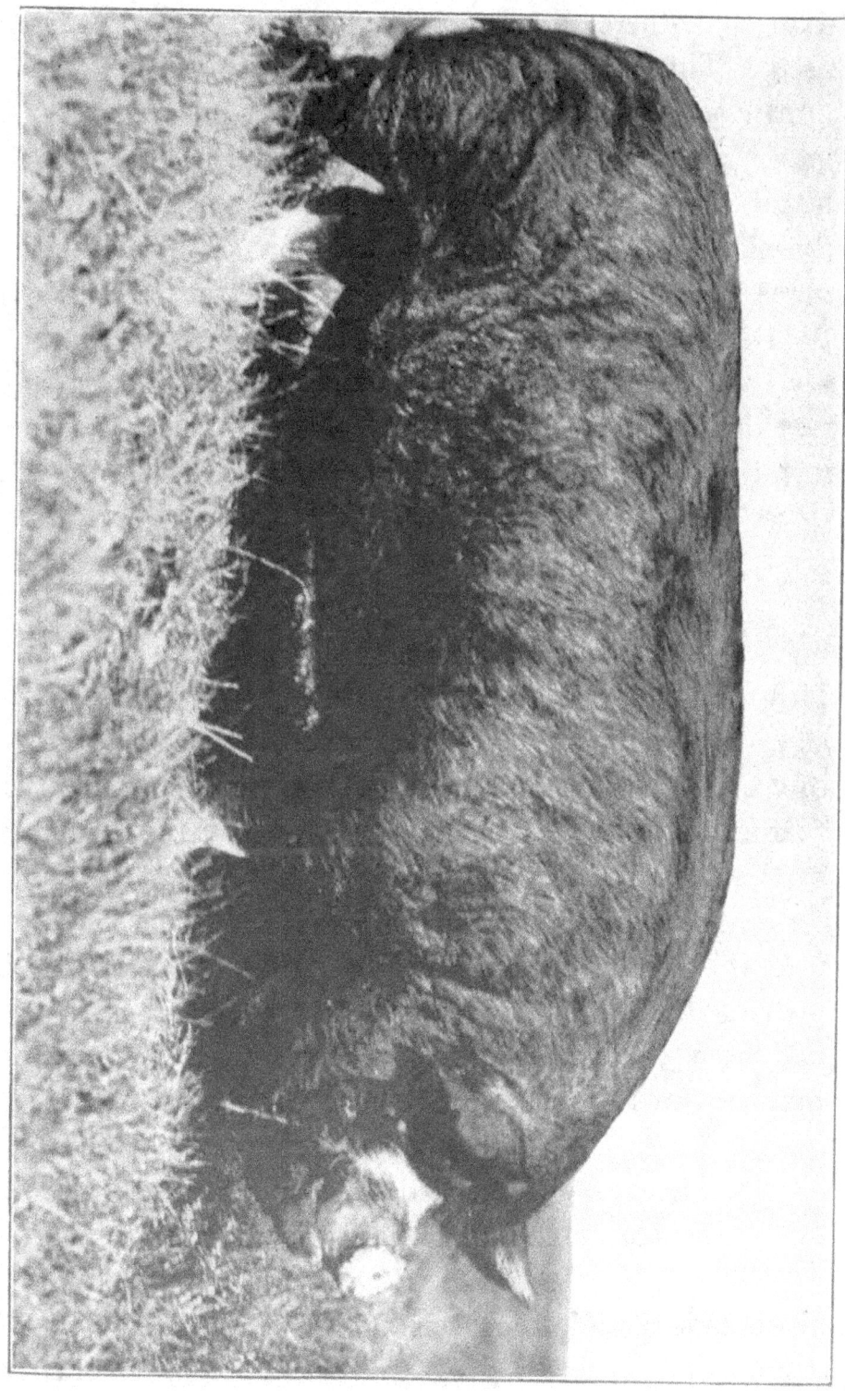

BERKSHIRE BOAR.

than the others did and the blood of such was freely used. This was no occasion to consider the idea of impairment of size or prolificacy. The great defects for many years were slow fattening and lack of market qualities. Until the beginning of the second quarter of the nineteenth century efforts to improve the swine were necessarily scattered and not very effective.

Developments of the years following gave promise of reward to breeders who could supply the most profitable type of swine to the rapidly increasing numbers of farmers in the corn-growing areas. The Berkshire was the most carefully bred hog obtainable, though his breeding in England had not been directed with a view to adaptation to utilization of corn. However he was superior in many ways to the native stock and gained a place. Selection among descendants of crosses of the Berkshire with stock combining the good features of the types previously used gave the foundation of the Poland-China.

Breed Building.

The pronounced disposition to fatten that characterized this breed brought it into strong demand for improving the stock upon farms where little improvement had been effected. The native sows being disposed to mature slowly but to reach good size and breed freely, the opposite extremes were really needed for mating with them in order that the offspring should be as nearly right as possible. Most of the breeders made their selections to meet the general demand. Heavy feeding of corn was commonly practiced.

Extremes Needed.

POLAND-CHINA BOAR.

Animals of the smaller size soon passed through the period of most rapid growth and became fat at an earlier age than those with greater tendencies to growth.

Because corn alone is more favorable to fattening than to growth its use in herds being bred for early-maturing qualities was an important factor in the elimination of animals that fattened less rapidly in their growing days. Well sustained gains are possible only by the continued development of frame that characterizes animals capable of coming to large size. The extreme of early fattening means rapid gains from accumulation of fat and also cessation of gains at a comparatively early age. It has been said that the tendency in all breeding of improved stock is to go to extremes. Extremes are usually demanded by stock-raisers who see the need of improvement in their previously neglected animals. Continued adherence to an extreme type in the herd eventually results in difficulty. If followed by a majority of those working with a breed it results in the stock becoming unsuitable to the needs of the raisers of commercial stock who first demanded the extreme type. This is no more true of swine than of other stock. Progress toward the extreme in swine was facilitated by the fact that corn-feeding aided the selection of less growthy swine. It also served to accentuate the tendency to small litters which naturally accompanies the curtailment of growth. Because one generation of swine follows another only twelve months later the results of selection appear in a very short time.

Results of Extremes.

Judges at fairs are usually breeders. They naturally

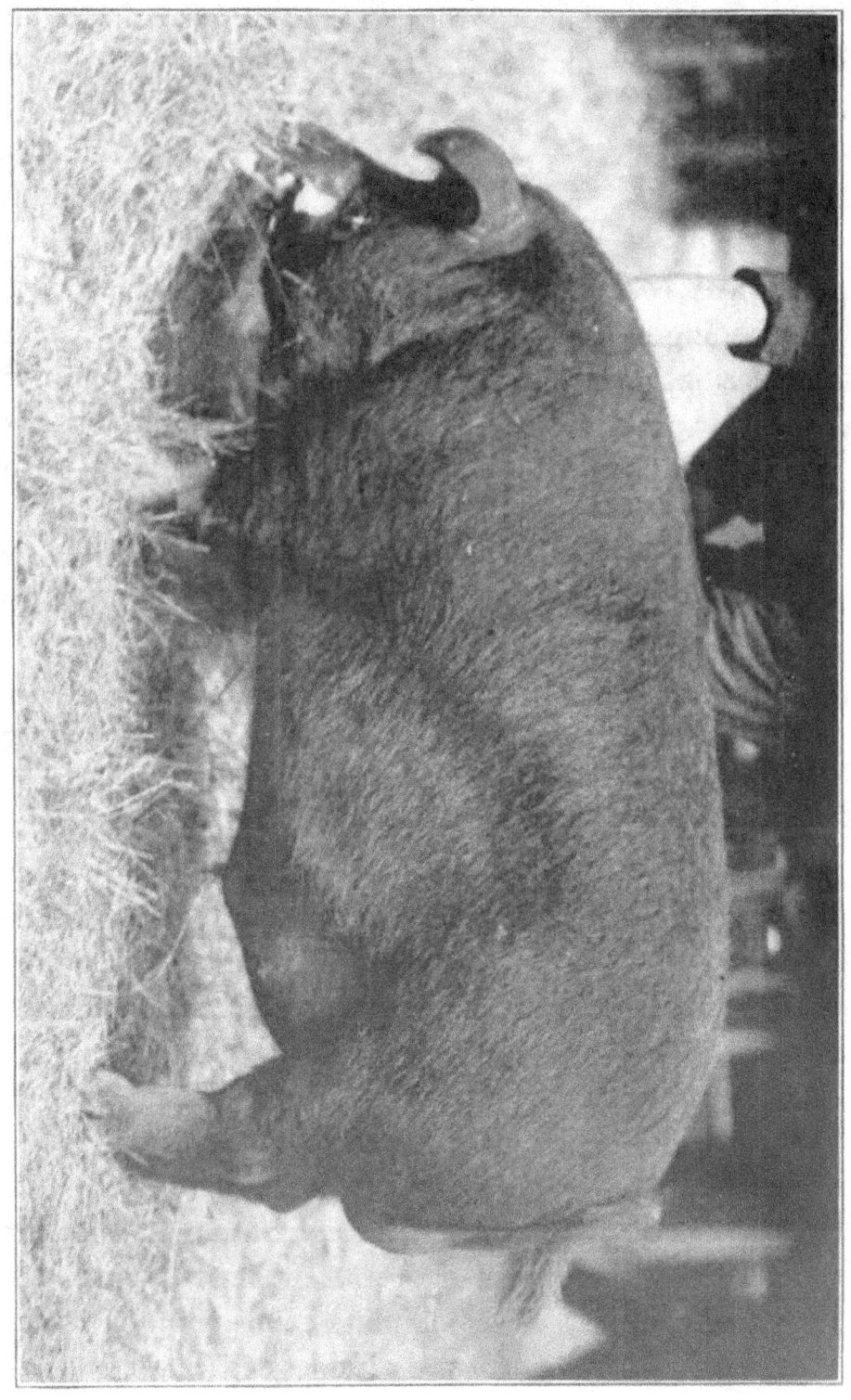

DUROC-JERSEY BOAR.

take as their standard the type of animal for which buyers will pay most liberally. The early type of show

Show Type.

hog in the cornbelt was the type that was demanded and needed by the buyers of sires of market hogs.

Developments in farm herds of swine also come rapidly and a comparatively few years of breeding to boars of the show type and of heavy corn-feeding brought the farm sows very close to the same type. It was then that complaints were made of lightness of bone, which means lack of size and growthiness. Smallness of litters was also an outcome of the same conditions and methods. The show type was spoken of as something separate and distinct from the farmers' type although it was first established as a result of farmers' demands.

The earliest improved breeds were the first to undergo such evolution in type and popularity. The logical remedy

Opening for More Breeds.

lay in the use of opposite extremes. These were usually found in newer breeds still retaining unimpaired size and fecundity along with native

coarseness and slowness in maturing. All the older breeds of swine have passed through much the same stage. The changes illustrate the idea that every wrong condition works its own remedy though after a great cost to individuals. The incoming breeds have not escaped the effects of the influences that brought them into demand though the workers with more recent introductions have seen the necessity of combining refinement and ready fattening qualities with a desirable degree of size and growthiness and have demonstrated the possibility of such a combination.

The rising and waning of the height of popularity of successive American breeds of swine compels one important conclusion. Neither real success nor profit can come to the breeder or raiser who proceeds by the mating of opposite extremes. Even though the mixing of breeds is avoided opposite types within a breed must necessarily be the product of different methods and the descendants of wholly different animals. The union of such is subject to the same uncertainties that follow blending the blood of animals of different breeds, even though the progeny remains eligible to registration. Errors having been made and realized, such a step may be the beginning of correction but it is a retracing rather than an advance.

Mixing Types.

Farm production of swine or other market stock is most satisfactory and most remunerative when the necessity for reversing methods and changing types or breeds is entirely avoided. If breeding stock is selected to embody the best possible combination of qualities for the market with the essentials of economic production, progress can continue without interruption. New sires selected for their possession and inheritance of the same qualities serve to raise the standard of the females and to reduce the proportion of inferior offspring by strengthening the blood. What is aimed at in bringing in new sires may also be contributed to in selection of the females which

Continuity in Farm Breeding.

become increasingly uniform and prepotent as time goes on.

There have not been wanting workers with the older breeds who foresaw the ultimate outcome of continued breeding in accordance with the extreme demands and needs of owners of wholly unimproved stock. Such far-seeing men have saved the day for their breeds by breeding the medium types which needed no correction and which they foresaw must ultimately be generally adopted. Occurrences in swine breeding also show the need of foresight on the part of professional breeders. To be permanently successful they must recognize and adhere to the essential points. The final result of the fads and extremes of transient popularity must be foreseen and avoided. The breeder has greater need than has the raiser of commercial stock to forecast demands that come through changing economical conditions or as a result of errors or misconceptions of the large number. It is with breeders endowed with such powers of perception that the permanency of our live stock industry rests. They have also frequent occasion to show the courage of their convictions by running counter to the ideas of a majority of their fellows or by giving the note of warning of the result of adherence to an impractical ideal.

Conservative Breeders.

Breeder's Reward.

The few breeders of unusual courage and judgment, upon whom so much rests at times, do not always live to see the return of their fellows to the conservative standards. The benefit of their work may sometimes go to those who follow them, though real ability as a breeder very seldom fails of receiving material compensation. The fascination of molding animal form makes the breeder's work an absorbing pleasure. To have earned the right to feel that he has aided in rendering domestic animals more useful to mankind is his most prized reward.

www.ingramcontent.com/pod-product-compliance
Lightning Source LLC
Chambersburg PA
CBHW081141180526
45170CB00006B/1886